CAMBRIDGE COUNTY GEOGRAPHIES

General Editor: F. H. H. GUILLEMARD, M.A., M.D.

BEDFORDSHIRE

Cambridge County Geographies

BEDFORDSHIRE

by

C. GORE CHAMBERS, M.A.
Late Assistant Master at Bedford Grammar School

With Maps, Diagrams and Illustrations

Cambridge :
at the University Press
1917

CAMBRIDGE UNIVERSITY PRESS

Cambridge, New York, Melbourne, Madrid, Cape Town,
Singapore, São Paulo, Delhi, Mexico City

Cambridge University Press
The Edinburgh Building, Cambridge CB2 8RU, UK

Published in the United States of America by Cambridge University Press, New York

www.cambridge.org
Information on this title: www.cambridge.org/9781107671942

First published 1917
First paperback edition 2013

A catalogue record for this publication is available from the British Library

ISBN 978-1-107-67194-2 Paperback

PREFACE

MOST people in Bedford, and not a few outside, knew the late Charles Gore Chambers as a man of great learning, wide reading, and deep thoughts. All his pupils found him sympathetic and ready to meet them on their own ground, whether language or literature, history or architecture, and he could draw out their ideas and suggest just the right thing to them. On almost any subject he went into details with loving thoroughness. And so, when he took this book in hand, though he already knew the county well, he set to work to explore it systematically, visited almost every place, entered every parish church, investigated both the past and the present state of the lace and straw industries, and in general put all his energies into the task that lay to hand. He was, in particular, much interested in the development of Luton and the past history of Dunstable; as regards the former he often dwelt on its continuous connection with Bedford, from the earliest days of which the Anglo-Saxon Chronicle has a record, down to modern days when the Midland

Railway connects them as once did the Alfred-Guthrum treaty line. I would also mention his keen pursuit of geological knowledge, especially his appreciation of the epoch-making discoveries of flints by Mr James Wyatt and Mr Worthington Smith, by which they established the existence of palaeolithic man.

It has been a great grief to us all that he was unable to correct the proofs. My task has simply been to make such alterations as he would certainly have made if he had had the chance (for the rapid lapse of time had already made some of his statements to be out-of-date) and to collect photographs. Mr W. N. Henman of Bedford had already taken some at his request, and we consulted together about taking or choosing others where we knew that he wished some place to be illustrated. Mr William Austin, of Rye Hill, kindly and readily provided some of Luton; and Mr Richmond, J.P., of Heath and Reach, did the same for Leighton.

J. E. MORRIS.

Bedford Grammar School,
March 1917.

CONTENTS

ILLUSTRATIONS

ILLUSTRATIONS

The illustrations on pp. 3, 14, 23, 25, 57, 61, 63, 69, 74, 88, 90, 97, 100, 102, 105, 107, 118, 120, 122, 137 and 157 are from photographs by Mr W. N. Henman; those on pp. 5, 12, 50, 71, 104, 112, 114, 124, 130, 141, 159, 174, 187 and 190 are from photographs by Dr J. E. Morris; those on pp. 17, 110, 123, 127, 128, 133, 135, 139, 142, 143, 148, 154, 165, 171, 176, 180 and 189 are from photographs by Messrs F. Frith and Co., Ltd.; those on pp. 31 and 36 from photographs by Mr G. G. C. Bull; the map on p. 39 is from the raised map by Mr F. Hawkins Piercy; the illustrations on pp. 66 and 181 are from photographs by Mr F. Thurston, Luton; the plan on p. 76 is by Mr A. R. Goddard; the illustration on p. 79 is from an educational wall card by Mr C. H. Ashdown; those on p. 106 from photographs by Mr S. Milne; those on pp. 116 and 179 from photographs by Mr R. Richmond; that on p. 156 from a photograph by Mr D. Macbeth; those on pp. 161 and 166 from photographs by Mr E. Walker; that on p. 186 from a photograph by the Author.

1. County and Shire. Origin of Bedfordshire.

England has been divided for at least nine hundred years into Shires or Counties. The word *shire* is the modern form of an old English word meaning a district or territory or department of administration—a part *shorn* off from a larger area. It has the same derivation as the word *share*, the verb to *shear*, and the *share* of a plough. At first shires were under the superintendence of an Ealdorman, a title later displaced by that of Eorl. But the connection between the Eorl (Earl) and the shire became less close, and the management of the shire passed into the hands of a Shire-reeve or Sheriff. This was already the case when the Domesday Book was compiled in the last quarter of the eleventh century. From the Conquest onwards official documents were for long written either in Latin or French. Earl appeared in Latin as *Comes*, in French as *Comte*; Sheriff as *Vice-comes* and *Vicomte*; while the Shire became *Comitatus* and *Comté*. But whereas Earl and Sheriff remained the English forms in general use, the French word *Comté* gave rise to the English *County*, which has survived as a familiar name for *Shire*.

A glance at a list of the names of the shires will disclose a difference of form that suggests a difference of origin. Many of them are simply formed by adding *shire* to the name of a town. If all such names be written down in their proper positions upon a blank map, it will be found that they lie continuously, and that some spaces south of the Thames, those along the coast between the Thames and the Wash, and the districts covered by Cumberland, Westmorland, Northumberland, and Durham, remain unoccupied. In other words the shires named from towns correspond to the old kingdoms of Mercia and Deira. The southern kingdom of Wessex gained supremacy over them early in the ninth century, but, before it had sufficiently established its power to organise the whole country under one central government, the Scandinavian invasions began. It was not till the middle of the tenth century that Mercia and South Northumbria were again restored to the West Saxon kings, and it is not till the latter end of that century that we find any mention of Midland shires.

Bedfordshire is first mentioned in the Chronicle under the date 1011. Bedford is earlier recorded as the place where the West Saxons gained a victory over the Britons at the end of the sixth century, and the towns of Hertford, Buckingham, Bedford, and Huntingdon are all mentioned as strongholds at the beginning of the tenth century; but there is no evidence that any regular provincial division existed at that time in this part of England. It will be clear then that there is a great difference between shires such as Essex and Kent and those like Bedfordshire or Cambridgeshire. The two former had gradually acquired a unity and form as separate kingdoms struggling with their neighbours and adapting themselves to the conveniences and defence of natural boundaries; whereas Bedfordshire and the counties which surround it would

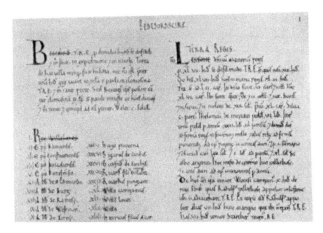

Part of the first page of Domesday Book, giving the entry for Bedford and the Royal land of Leighton

seem to have been carved out to meet the requirements of the collection of taxes, the administration of justice, and the organisation of a militia. It is impossible to do more than conjecture the causes which led to the great difference in their sizes, and the variety and irregularity of their shapes. While such counties as Kent and Essex are older than their towns, the town of Bedford is far older than the shire; and whereas Chelmsford and Maidstone are towns in Essex and Kent, Bedfordshire is rather a shire about Bedford.

It seems likely that the following were some of the conditions under which Bedfordshire assumed the form which it practically retains at the present day. Bedford was to be its centre and citadel. To suit the arrangements of the Exchequer, the county was to contain about 1200 hides, the hide representing the unit by which land taxes were assessed. It was not advisable to divide big parishes such as those of Luton and Eaton Socon; and if possible a county should be wholly within one diocese. The county boundary would often be drawn irregularly to include the whole of a manor instead of dividing it between two shires. Physical conditions do not appear to have had much influence in shaping Bedfordshire.

2. General Characteristics. Position and Natural Conditions.

The natural features and position of a county, as we shall see, have no little effect upon its history. The earliest evidences of human occupation in Bedfordshire are to be found, as might be expected, upon the higher and drier ground, upon the gravel banks that had accumulated in the bed of the prehistoric Ouse and elsewhere, and along the elevated chalk tracts. The settlement of England by Saxons and Angles was effected by tribes which landed some on the south, some on the east coast, and only gradually made their way into the interior. The position of Bedfordshire makes it likely that for a long time much of it was a "no man's land." It was probably reached from the Wash by the waterway of the Ouse, and from East Anglia by the chalk hills

The Greensand Range seen from Cleat Hill

along the south, but there is no record of any such invasion. The West Saxons undoubtedly approached it along the chalk ridge escarpment from the south, following the line of the Icknield Way from the Thames, and in the seventh century it was included in the Anglian kingdom of Mercia. But before Anglo-Saxon times the

Roman occupation of Britain had made systematic communication with the north and west necessary for London as a trading centre and Colchester as a military centre. The main military road from Colchester to the north (the Ermine Street) had to avoid the fens, and passed through Cambridgeshire at no great distance east of Bedfordshire, while an alternative length of it followed the Ivel through Biggleswade and Sandy. The great north-western route (the Watling Street) only crossed the south-western corner of Bedfordshire at Dunstable. The position of these roads thus left the county somewhat in the lurch. The south-west, south, and south-east parts were in easy communication with the outer world, while Bedford had its water road, the Ouse, and was probably linked up from early times by a roadway over the gravel to Sandy. But the clay hills of the north, the marshy plain and sand deserts of the centre, and the bogs of Flitwick and Westoning remained isolated.

Indirectly, too, the position and physical features of the county affected its history in the ninth century. When Wessex made terms with the Danes the Ouse above Bedford, a line from Bedford to the source of the Lea, and the Lea itself were selected as a frontier. The reason for taking this line, instead of the more direct line of Watling Street, was doubtless to include Luton and the south-western part of Bedfordshire, which had been originally taken from the Britons in the sixth century, within the area of Wessex.

In prehistoric days scrubby forest covered much of

what was not marsh. When the woods were cleared from the clay lands which occupy two-thirds of the county, the soil proved well adapted for the growth of wheat, and though the central belt of sand was not valuable for the same purpose, its intermixture with the clays upon its slopes and along the valley of the Ivel produces a loam of exceptional fertility. On the chalk downs of the south nature afforded abundant food for sheep, and in the rich meadows that fringe the Ouse and the Ivel there was excellent pasture for cattle. Thus Bedfordshire was naturally an agricultural county, and the produce of the whole area was in easy reach of the metropolis. Wheat and barley were the staple products, and in course of time, as has happened in the case of other industries, one of these by-products grew in importance. The plaiting of straw and the manufacture of hats became specialised, and centred at the grain markets of South Bedfordshire and North Bedfordshire. It was probably its position upon Watling Street that gave Dunstable the precedence, and competition, helped by the extension of the Midland Railway to London, that carried the trade to Luton. The greensand hills from Leighton to Potton were for ages little more than barren waste covered with woods. But it was a hunting ground, and as such was constantly visited even in the sixteenth and seventeenth centuries by Henry VIII and James I. The Cistercians, who always sought the waste land and made efforts to turn deserts into gardens, had settled here at Woburn and Warden in the twelfth century, and in course of time a succession of parks and country seats spread over the range and its slopes.

The nearness of Bedfordshire to London had another result. Every progress of a monarch, every expedition to the north-west or north taken by judges or adminis-trators made the county known, and many government officials, successful lawyers, and wealthy London merchants bought estates and settled in the county. On the other hand many Bedfordians successfully sought their fortunes in London and brought wealth into the shire. In recent years, too, the position of Bedfordshire has brought great changes. When the Midland Railway sought fresh fields for its energies by securing its extension to London, it found the two ancient routes of Watling Street and Ermine Street occupied by the London and North-Western and the Great Northern Railways. It struck a line between them, and so opened a new era in Bedfordshire, as it not only enormously developed the trade of Luton and the educational conveniences of Bedford, but brought to both a number of important manufactures. Yet the northern upland of the county has still remained untouched.

3. Size. Shape. Boundaries.

Although the demarcation of the county boundary does not as a rule follow marked physical lines, it approximates to the line of watersheds on the extreme east and on the north-west, and on the west it roughly follows the minor water-parting between the main Ouse and its tributary the Ousel or Lovet, which stream forms the county limit for a short distance between Eaton Bray and Heath, a northern member of Leighton Buzzard. The Ouse itself is the boundary for a few miles north and south of Turvey, and again in the north-east from north of St Neots to south of Eaton Socon; while the boundary follows the Rhee, a tributary of the Cam, for a mile or two south of Dunton. The shape of the shire is an irregular oblong, lying north and south, and bulging to the east into Cambridgeshire. Its greatest length from north to south is 36 miles, its greatest breadth east and west 20 miles; but below the eastern bulge it measures only about 12 miles across, and this again is reduced to about seven in the chalk extension south of Dunstable and Luton. The most marked irregularities in its bounds, which had remained with little alteration from the time of the Domesday survey, were rectified by the Local Government Act of 1888. The parish of Swineshead, though nearly shut off from Huntingdonshire, belonged to that county, but is now in Bedfordshire, and Tillbrook has been surrendered in compensation. A spur of Hertfordshire used to intrude between Luton on the east and Studham and Whipsnade on the west, with the result that Kensworth, Caddington, Studham, and Markyate Street were divided between the two counties; the three former are now wholly in Bedfordshire, and the last belongs to Hertfordshire. At Everton on the north-east there was a complicated threefold division between Huntingdonshire,

Cambridgeshire, and Bedfordshire; it is now for most purposes in Bedfordshire, but Tetworth, which forms ecclesiastically a joint parish with it, is in Huntingdonshire. An island of Hertfordshire containing the Chapel Farm and the old Chapel of St Thomas lay in Meppershall parish; but it is now included in Bedfordshire. The explanation of such irregularities often lies in claims of manorial jurisdiction over outlying members of a manor, the outlying members of a manor in another county being returned as part of such manor in the county to which it belonged.

Bedfordshire is smaller than any other county with the exception of Rutland, Middlesex, and Huntingdonshire. Its area measures 303,000 acres, that of Yorkshire nearly 3,000,000, that of Rutland less than 100,000. But it will be perhaps more useful to compare it with the group of counties which surround it. Of these Huntingdonshire alone is smaller, having rather more than two-thirds the acreage. Bedfordshire is equal in area to about three-quarters of Hertfordshire, rather more than three-fifths of Buckinghamshire, rather more than half of Cambridgeshire, and less than a half of Northamptonshire.

4. Surface and General Features.

The north of Bedfordshire rises almost immediately from the north bank of the Ouse into a mass of clay divided into separate blocks by the brooks which have marked out valleys or denes. It is composed of boulder-clay on Oxford clay, and affords a rather monotonous scenery chiefly consisting of cornland with little wood. The least inviting part is that which stretches from Knotting past Bolnhurst and Keysoe to Eaton Socon. The descent to the north towards Kimbolton is picturesque, the view down over the Ouse from Tempsford to St Neots as seen from the road between Barford and Eaton Socon is a fine one, while in the west there is much picturesque broken ground about the brook that passes through Sharnbrook. The north-western rectangle lying north of a line from Sharnbrook to Harrold is largely under grass, and in many respects seems to belong rather to Northampton: most of it indeed drains to the Nene. Along the whole course of the Ouse, and to some extent along the Ivel, a succession of rich grass meadows and endless willows and alders afford the quiet charm characteristic of sluggish southern English rivers. South of the Ouse the Vale of Bedford lies flat and uninteresting for a breadth of almost five miles. Passing travellers often identify Bedfordshire with this kind of scene, but it is really not characteristic of the county. South of the plain of Bedford, in a curve sweeping from Leighton Buzzard on the south-west to Sandy and Potton on the north-east and broadening over a width of about five miles, is the range of the lower greensand. It is a region of parks and plantations and reclaimed waste and bog. On the lower edges of it to the east and south, where the surrounding clay intermingles with the sand, its rich loam gives the opportunity for much vegetable and flower gardening, and Sandy and Biggleswade send large supplies of such produce to the London and northern markets. Aspley Guise and Woburn Sands are notable as health and holiday resorts, their elevation, sandy soil, and pine woods adding to the attractions of charming and varied scenery.

Sharpenhoe Chalk Heights seen from Toddington

South of the sand range there is a narrow low-lying track of gault clay, which passes upwards into chalk marl, and so rises on the south in the fine chalk escarpment that is so conspicuous at Sharpenhoe and Barton. The chalk downs gradually rise to about 540 feet above Luton, to about 600 above Dunstable, and to 800 at Kensworth three miles farther south.

The county thus shows four parallel belts, the northernmost reminiscent of its neighbour Northamptonshire, the Vale of Bedford suggesting Huntingdonshire, the greensand range in the middle which recalls Surrey, and the Chiltern District of the south that might be easily matched in Berkshire or Wiltshire.

5.   Watersheds.   Rivers.

Water which falls upon the surface of Bedfordshire may reach the sea by several channels, the Nene, the Ouse, or the Thames.   It may reach the Ouse directly, or by the Ivel, within the county; by the Ousel (or Lovet) at Newport Pagnell in Buckinghamshire; by the Kim, which is mainly in Huntingdonshire; or by the Rhee and Cam in Cambridgeshire.   It may flow into the Thames between Runnymead and Staines, passing thither by the Ver and Coln, or it may be borne by the Lea into the lower reaches at Blackwall.   Of these the Rhee takes so little water from Bedfordshire as to be negligeable.   On the extreme north-west the watershed runs south-east from the neighbourhood of Souldrop along the "forty-foot" (a disused grass track) between Hinwick Lodge and Colworth House, across Odell Wood to Dungee Corner at the north of Harrold Park.   From the northern part of Odell Wood a stream runs north-west through Hinwick, joins another which rises north of the forty-foot and east of Hinwick Lodge, passes Poddington, and, reinforced by a brook from Wymington, leaves the county to the east of Farndish and makes its way into the Nene.

The Ver leaves the base of the upper chalk in the depression up which the Watling Street passes between Caddington and Kensworth; the Ousel rises at Wellhead on the Icknield Way, where it skirts the western

Source of the River Lea at Leagrave

foot of the Dunstable downs; and the Lea rises about two miles north of Luton and three east by north of Dunstable, where several springs bubble up close to Leagrave railway station. These three streams are thrown out at the base of the lower chalk by the hard layer of Totternhoe stone, which arrests the further penetration of the water stored in the huge mass of lower,

middle, and upper chalk above.   No part of the surface
of these downs retains water except a few hollows that
still hold vestiges of drift or other clays, and these are
merely stagnant ponds of no great depth.   At Dunstable,
which is 480 feet above mean sea-level, wells must be
sunk for nearly 80 feet to obtain water; those sunk on
Caddington and Kensworth hills are proportionately
deeper; indeed a well at Mount Pleasant, which stands
but 40 feet below the highest point of Kensworth hill
(800 feet), is 350 feet in depth[1].   Four hundred feet
above sea-level is the average height at which springs
flow from the base of the lower chalk, the Ousel rising
at 420, the Ver at 434, and the Lea at 370.

The northern escarpment of the lower chalk runs
slightly south of west from Barton by Sharpenhoe,
Charlton Cross, and Houghton Regis to Billington, and
completes the water-parting between the Ver and Lea
on the south, the Ousel on the west, and the catchment
basin of the Ivel on the north-east.   Turning at an
angle to the north-west from Charlton Cross, a line
passing along the gault on the south of Toddington
Hill and crossing the sand range by Eversholt and the
east of Woburn Park, reaches Ridgmont, sends down
Crawley Brook on the west to the Ousel, and on the
east two streams from Ridgmont and Toddington Hill.
These both flow into Flitwick Bog, and, leaving it as
the Flitt, pass Clophill, and are joined at Shefford by
a stream from the south that has come from below

---

[1] Several of these deep wells are worked by donkeys, which stand
inside a hollow cylinder and tramp it round as a squirrel turns his cage.

Harlington, passed south of Wrest Park and fed its ornamental waters, and been reinforced by another from Barton. The Flitt runs on from Shefford to the Ivel, which it joins above Langford.

The Ivel itself rises near Baldock in Hertfordshire, enters the county between Stotfold and Astwick, and receives the water of the Hiz just north of Arlesey. On the east it is divided from the Rhee and Cam by the clay heights of Dunton. At Lower Caldecote it receives a stream that comes down from the clays of Cambridge-shire to the east of Gamlingay, by Potton and Sutton, fed by two branches, one from Wrestlingworth and one that comes to it through Stratton Park from the neighbour-hood of Edworth. The only western tributary of the Ivel is a small brook that joins it at Girtford.

North of the Ouse the water-parting is the high ground lying between Thurleigh and Bolnhurst on the south, and Keysoe and Souldrop on the north. The brooks that run north join the Kim and eventually reach the Ouse near Hale Bridge, north of Eaton Socon; to the south and east others run direct into the Ouse.

The sand range from Southill to Millbrook parts the Flitt and its affluents on the south from the brooks that flow directly into the Ouse on the north, but supplies little water on either side. The brooks to the north drain little more than the surface clay. West of Millbrook the elevated line of the sand continues in a curve of clay hills from Ridgmont to Brogborough west and then north to Cranfield. From the high ground south-west of Stagsden two streams come down from

Astwood and North Crawley (both in Bucks), and uniting just beyond Stagsden, have cut a deep bed of nearly a mile in width into the oolite from their junction to the Ouse at Bromham. From the line between Souldrop and Odell Wood, which parts the Ouse from the Nene, a brook runs into the Ouse at Sharnbrook. This stream too, though its course is short, has cut both wide and

The Ouse at Bedford

deep through the cornbrash into the lower oolite. The oolite is full of fissures, and so forms a natural system of underground pipes and reservoirs by which great quantities of water pass unseen in a south-easterly direction. Bedford is supplied with water by wells which are extended by headings driven across the strike of this underground river. In the neighbourhood of the town it maintains a level which is approximately that of the Ouse with which it is in connection, about ninety feet above mean sea-level. A glance at the map will show that the sand-hills of the centre supply very little water to the brooks and rivers. The sand, which is of considerable depth and extends beneath the gault at least as far south as Hitchin, is a great reservoir of water. Where it is underlain by clay, as for instance at the brick-pits to the north of Sandy, it throws out springs at its base. But the greensand range of central Bedfordshire, though resting largely on Ampthill or Oxford clay, is to a great extent sealed up by gault or boulder clay that banks its slopes on south and north. The Biggleswade waterworks draw their water from a mass of greensand that lies between a clay bottom and a gault clay covering: the latter protects the area from all surface pollution, and the water pumped from it is that which has fallen as rain on the sand range to the north and west and passed through the natural filter of at least some few miles of sand. Many of the neighbouring villages are supplied by Biggleswade, even so far afield as Kempston. Leighton Buzzard also draws its water from the greensand, while Dunstable and Luton depend upon the chalk.

There are no lakes in Bedfordshire, but in the flat bottom south of Ampthill where the clay partly overrides the sand there are still considerable remains of Flitwick moor and bog. More than a hundred years ago efforts were made to drain it. Of late years considerable advance has been made, and between Greenfield and

Flitwick beds of vegetables and fruit bushes may be seen in luxuriant growth within a few yards of the retreating bog. Its disappearance will be the gain of the small cultivator and the grief of the naturalist, for it harbours many interesting (and perhaps some unknown) species of plant and animal life.

6. Geology.

Bedfordshire is so small a county that it might be expected to offer scant space for geological variety. But within its narrow bounds there is more than any one of the five surrounding counties can boast. Its surface displays the successive strata of the Jurassic and Cretaceous series with few exceptions, and the explanation of the absence of those not represented makes the study of its formation still more interesting and suggestive.

Of the Tertiary rocks there is but little trace. If they were ever deposited over more than a very small part of the south of the county they have been long since washed away and have left but negligeable remains. Of the Glacial period that followed there are abundant evidences. By ploughing and pounding, by tearing and crushing, by depositing huge masses of clay and stones upon the surface, the glaciers of the Post-tertiary period have set their mark on every corner of the county.

Names of Systems		Subdivisions	Characters of Rock
TERTIARY	**Recent Pleistocene**	Metal Age Deposits Neolithic ,, Palaeolithic ,, Glacial ,,	Superficial Deposits
	Pliocene	Cromer Series Weybourne Crag Chillesford and Norwich Crags Red and Walton Crags Coralline Crag	Sands chiefly
	Miocene	Absent from Britain	
	Eocene	Fluviomarine Beds of Hampshire Bagshot Beds London Clay Oldhaven Beds, Woolwich and Reading Thanet Sands [Groups	Clays and Sands chiefly
SECONDARY	**Cretaceous**	Chalk Upper Greensand and Gault Lower Greensand Weald Clay Hastings Sands	Chalk at top Sandstones, Mud and Clays below
	Jurassic	Purbeck Beds Portland Beds Kimmeridge Clay Corallian Beds Oxford Clay and Kellaways Rock Cornbrash Forest Marble Great Oolite with Stonesfield Slate Inferior Oolite Lias—Upper, Middle, and Lower	Shales, Sandstones and Oolitic Limestones
	Triassic	Rhaetic Keuper Marls Keuper Sandstone Upper Bunter Sandstone Bunter Pebble Beds Lower Bunter Sandstone	Red Sandstones and Marls, Gypsum and Salt
PRIMARY	**Permian**	Magnesian Limestone and Sandstone Marl Slate Lower Permian Sandstone	Red Sandstones and Magnesian Limestone
	Carboniferous	Coal Measures Millstone Grit Mountain Limestone Basal Carboniferous Rocks	Sandstones, Shales and Coals at top Sandstones in middle Limestone and Shales below
	Devonian	Upper Mid Devonian and Old Red Sand- Lower stone	Red Sandstones, Shales, Slates and Lime- stones
	Silurian	Ludlow Beds Wenlock Beds Llandovery Beds	Sandstones, Shales and Thin Limestones
	Ordovician	Caradoc Beds Llandeilo Beds Arenig Beds	Shales, Slates, Sandstones and Thin Limestones
	Cambrian	Tremadoc Slates Lingula Flags Menevian Beds Harlech Grits and Llanberis Slates	Slates and Sandstones
	Pre-Cambrian	No definite classification yet made	Sandstones, Slates and Volcanic Rocks

In the far distant ages of the Triassic period, long before Great Britain had assumed an island shape, there stretched over it a great inland salt-lake as far as the coast of Normandy. It was narrowest at its central part, between Bedfordshire and the Mendip hills, and, while the Welsh mountains rose on its north-western shores, its south-eastern coast trended from the Wash south-westwards, leaving Bedfordshire and the land east, west, and south of it as a dry highland of rocks. The soil produced ferns and cycads, equisetums—huge plants like our modern horsetails—and conifers. Crocodiles inhabited the waters, and, amongst Dinosaurs and other lizard-like monsters, one kind at least of marsupial had already made its appearance. The climate was warm and dry, there was little rain, and rock-salt accumulated beneath the water of the shrinking lake. Streams were however pouring the waste of surrounding hills into the lake, and the muddy scour of the coal measures from the west and north was laid down in beds of that black shaly Lias that crosses England from Gloucestershire to Yorkshire, and is so familiar to those who have visited Whitby. This Lias bed encroached upon the sinking north-west corner of our area, and, though it nowhere crops out upon the surface, underlies the upper rocks about Felmersham and Sharnbrook, and as far east as Bedford.

As time went on, this lake, by the sinking of intervening land, became connected with a southern ocean. The inflowing waters altered the conditions of life: greater evaporation brought more rain, and consequently a more abundant vegetable and animal life. The inland lake had now become an arm of the sea, which covered the north-western corner of the land corresponding to our county; its warmer waters bore other kinds of living creatures, and sand and limestone were deposited about its shores in beds that no longer took their colour from the waste of coal-fields, while in water of convenient depth coral reefs were built up. The coast meanwhile receded slowly to the south-east as the land sank. The "Northampton sands" lie beneath Wymington, and a succession of sand and sandy clay and limestone known as the Upper Estuarine Series underlies Bedford and the neighbourhood at a depth of seventy feet, and comes to the surface near Harrold and Sharnbrook. Much of northern and north-western Bedfordshire was now beneath the water. As the shore continued to recede and the sea to advance, successive beds of clay, marl, and limestone were deposited, the Oolite of to-day, which lies beneath the district that is roughly contained between Poddington, Stagsden, and Bedford; and the Ouse has worn its channel deeply into it between Turvey and Bedford. How much further south or east it extends is quite uncertain, as it is deeply buried by overlying deposits. Above it was formed a shallow layer of the greyish rubbly limestone known as Cornbrash, which is exposed at intervals along the Ouse valley above Bedford and in the two old brooks that join it at Bromham and Sharnbrook, and may also be seen in the large Biddenham gravel-pit. Its fossils show that it was formed at some distance from land, and it is probable

that when it was deposited most part, or all, of Bedford
was already submerged. Above it, but separated by a
layer of clay, lies a limy sand called Kellaways Rock,
whose fossils tell us that it was laid down in shallowish
water, and it probably marks an alternative period of
rise in the sinking sea floor. It is a thin bed, containing

"Doggers" in Kellaways Rock

iron pyrites and concretions of sandstone called "doggers," well seen in the Oakley railway cutting before the
banks were overgrown. The Cornbrash and Kellaways
Rock are overlain by a thick bed of Oxford clay, forming
strata of dark blue-grey and brownish clays alternating
with thick beds of limestone. As some five hundred

feet of it were deposited, the sea must have been of considerable depth, and it undoubtedly extended far beyond Bedfordshire on every side.

After ages of gradual sinking, during which this 500 feet of Oxford clay was deposited, the land began to rise again, and above the Oxford clay a stratum known as the Corallian formed, but there is no trace of coral-reefs having existed within our area and our representative of the Corallian is a clay that differs but little from the Oxford clay beneath it, and is known as Ampthill clay, because it is well exposed in the railway cutting near Ampthill station. Above it a bed of Kimmeridge clay once lay over part at least of the county, but was stripped from its surface, and the only evidence of its presence that we now possess consists in those water-worn concretions, the red coprolites of the Greensand, the nucleus of which is often a fossil from the Kimmeridge clay. Much of the Ampthill clay must have been also denuded, for at Sandy, on the axis of the highest elevation, the Greensand rests directly upon Oxford clay. And there is little doubt that some of the upper levels of the Oxford clay were washed away as well.

The Greensand was now deposited. Much of Bedfordshire had become dry land between two seas, and now a time came when their waters began to overwhelm the intervening isthmus. The two advancing gulfs eventually met, and, as the channel broadened, the sea laid down along its coast a belt of coarse sands—the Greensand—which began at no great distance to the north-west, and gradually extended south-east as far as the neighbourhood of Hitchin. As the land sank, the sea widened and deepened, and the deposits, now farther from the shore and of finer materials, formed the bed of Gault which still overlies the Greensand in South Bedfordshire, and once extended to the north over that which is now exposed. It took its dark-bluish tint

Sand-pit at Cainhoe near Clophill, showing the
infiltration of Peroxide of Iron in bands

from the waste of Oxford clay. Again the character of the sediment gradually changed, and the Gault was overlaid with beds of chalky marl, in which the waste of the land brought down by rivers mingled with a gradually preponderating volume of waste from the living organisms that inhabited the water, in its turn succeeded by the Lower Chalk, which rests upon a thin band of hard brownish stone with phosphatic nodules at its base, quarried for a very long period at Totternhoe, and known as "Totternhoe Stone." Above a hard stratum, the Melburne Rock, is the Middle Chalk, which has also a band of hard chalk above it, and the presence of these strata of harder rock which occur at intervals is due to changes brought about periodically by varieties of temperature, due either to a widespread alteration of seasons, or more probably to the influence of colder currents that persisted for a time and then altered their course. Sufficient variation in the temperature of the water would affect for a time the character of the organisms living in it and of the débris precipitated to the bottom. The Upper Chalk, which is 800 feet thick at Kensworth, was once much thicker, and lay many fathoms beneath the ocean which stretched from the borders of the Welsh mountains and over much of what is now the continent of Europe.

For a vast period of time, a period that we cannot at present calculate, Bedfordshire was buried beneath this ocean, and with it all relics of its earlier types of animals and vegetation. The waters gradually subsided, and the land rose again. Races of animals and plants were developed and transformed: countless kinds, many of which, doubtless, have left no record that we have yet discovered, took shape and flourished and passed away, until at last the fauna and flora of our globe became such or nearly such as we know it now. But we cannot trace the history of the land through these long ages. The depression that again plunged the London

Basin beneath the waters and added fresh deposits to its surface has left some trace in isolated patches of Tertiary clay at Caddington and Kensworth, but we do not know how much of the area of the county was submerged. It is clear that during this long space of time the agencies of water, frost, and wind, must have remodelled the surface of the land. Great rivers tore up and washed away the Chalk and much of the Greensand and Gault, and left the denuded clay plain of Bedford bordered by the escarpment of this Greensand, and the Gault by the Chalk escarpment, both of which stood probably but little in advance of their present positions. The general configuration of the land as it then was would of course more closely resemble its present form if the Boulder Clay could be removed from the surface.

The next great change was brought about by a revolution in climate. The temperature gradually fell, and arctic conditions extended over much of Europe. Huge glaciers crept down from the north and north-east, from North Wales, from either side of the Pennine range, and from the neighbourhood of Norway. From the absence of hard rocky soil it is not as easy to trace the exact course of the ice here as in some districts of the north, but one glacier appears to have made its way by Leicester and Buckingham towards Aylesbury, while another entered Bedfordshire from Lincolnshire by the Cambridgeshire border, and passed over the eastern and southern parts of the county, detached lobes of it probably penetrating the several spaces between the masses of higher land. These pushed before and beneath them, or carried upon their surfaces, the wreckage of the havoc they had caused elsewhere. Much of this they deposited in their passage, some of it was laid down in stratified beds by the water that left the glaciers with the recurrence of the warmer seasons; and everywhere masses would be left in chaotic deposit when the glaciers finally melted. From one end of the county to the other was spread an almost continuous sheet of chalk and clay and stones—the Drift. Chalk forms the most conspicuous ingredient of the Drift clay and was in part torn from our own chalk hills, in part borne from the northern Midlands and the Lincolnshire wolds. The melting ice released a volume of water that must have washed out much of the chalk, and left beds of boulders and gravel; and the gravels were in turn driven along, re-arranged, and deposited by the rivers in and along their beds, as the inclination of the land led the escaping water along its path of least resistance. Before long a regular drainage system took shape. The Ouse still flows in a narrow channel that winds over the floor of the bed once occupied by a mightier stream, which made itself a path two miles in width through the Boulder Clay. In course of time its narrowing channel deepened, as it cut so deeply through the underlying strata that the dwindling stream now runs from Turvey to Bedford upon the oolite rock at a height of not much more than seventy feet above sea-level. The Ivel followed a course that an earlier stream had excavated, a deep channel filled with Boulder Clay, through the upper part of which the modern river runs.

7. Natural History.

The British Isles are, in geological language, of the type known as Recent Continental Islands. They formed at one time part of the large continental mass of land lying to the south-east, and have not been separated from it for any very great length of time, geologically speaking. The coast-line in many parts of the world— ours among the number—is constantly altering; in some places, as on the Norfolk coast, being ceaselessly eaten away by the sea; in others extending its limits to a not less remarkable degree. Between England and the continent the sea is very shallow, and even in the North Sea, if St Paul's Cathedral could be placed in it, there is no spot where its summit would not be above the surface of the water. But a little west of Ireland we soon come on to very deep soundings, which mark the original limit of the continent. To geologists these submarine evidences of a former land connection are known as the Continental Shelf.

Originally, no doubt, the fauna and flora of our land were the same as those of the continent of which it formed part. But at one period the islands were almost entirely submerged, and their animal and vegetable life being thus destroyed they would have to be re-stocked from the continent when the land again rose, the influx of course coming from the east and south. Later, separation once more occurred, before all the con- tinental species had had time to establish themselves on our land. We should thus expect to find that the parts in the neighbourhood of the continent were richer in species and those farthest off poorer, and this proves to be the case both with plants and animals. While Britain has fewer species than France or Belgium, Ireland has still less than Britain.

Apart from this factor the richness or poverty of a district is dependent primarily upon geological and physical conditions. Certain soils are favourable for the growth of certain trees or plants, which afford the food of special insects or animals, in their turn the prey of other creatures. The more varied, therefore, the geological conditions of a county, the richer probably will be its fauna and flora, and if it can show mountain and plain, moorland and river valley, forest and fen, sea-coast and lake, it will afford the naturalist a much longer list of species than if more restricted in its physical features.

Let us now turn to Bedfordshire and see what it has to offer. To begin with, it is a small county—one of the very smallest, Hunts, Middlesex, and Rutland alone being of lesser area. It is an inland county, has few marshes, no fen-land, no lakes worth mentioning, and but little limestone. From early times it has been highly cultivated and much drained, and this has no doubt helped to kill out many of the native species. The highest point is only about 800 feet, and though the northern part of the county is hilly it is nowhere much above 300 feet.

These facts, then, prepare us for a certain poverty in

vegetable and animal life, and though this is perhaps partly counterbalanced by the county being situated on the line of migration of birds and by the existence, in the central and less cultivated part, of a number of

Nest of Tufted Duck

parks, in which many animals and plants are protected, whether accidentally or purposely, this is, in fact, what we find; though it is possible that this poverty may to a certain extent be apparent rather than actual, due to the small attention that has been given to the county by naturalists.

As already stated, cultivation and drainage have doubtless much lessened the flora. Over a great part of the parishes of Ampthill, Maulden, Westoning, and Flitwick, spread, not very long ago, a waste of moor and bog, the disappearance of which was largely due to the more thorough and careful agriculture which began to prevail towards the end of the eighteenth century. But to this day the remains, or even the dry sites of these, afford good hunting-grounds to the botanist; such places, for instance, as Flitwick, Priestley bog, Westoning moor, Sutton fen, Gravenhurst moor, and the marshy lands at Potton, Ampthill, Stevington, Cainhoe, and Clophill. Besides the less common sedges and water-plants, they still show the grass of Parnassus, the sundews (*Drosera*), water violet (*Hottonia palustris*), bog-asphodel (*Narthecium ossifragum*), buckbean (*Menyanthes trifoliata*), great spear-wort (*Ranunculus lingua*), and marsh orchid (*O. latifolia*). The arrowgrass (*Triglochin palustre*) still grows on the site of the vanished Graven-hurst moor, and the sedge (*Carex pulicaris*) occurs with grass of Parnassus on what is now the dry chalk of Markham hills. So, too, the sedge on the lower slopes of Cleat Hill tells us that the scanty brook below must have been a fuller stream at the time of Domesday Book. The ditches yield bladderwort (*Utricularia*) and the pretty willow-weed, and in one pond at least the beautiful Villarsia, somewhat of a rarity, has been found.

Many of the older woods show plants of interest—the wild teasel (*Dipsacus sylvestris*), herb-paris, Solomon's seal, meadow saffron, and fritillary. Putnoe Wood contains the water-avens (*Geum rivale*) and the lily of the valley is to be found in more than one locality.

But perhaps the most characteristic botanical feature of Bedfordshire is the prolific plant life of the Ouse and the Ivel. The placid streams are starred for miles in summer with the white and yellow water lilies, and their banks lined by thick beds of sedge and feathery reeds. Here we see the spikes of the sweet-flag (*Acorus calamus*) and the rosy umbels of the flowering rush (*Juncus floridus*) interspersed with fields of sky-blue forget-me-not, while wide stretches of the stream are crimson with persicaria. The beautiful blue flowers of the meadow crane's-bill are too seldom seen, but the willow-herb and yellow flag are abundant. These are only some of the many plants which combine to form the characteristic river-scenery of our county.

The clay, chalk, and Greensand soils exhibit the plants specially affecting them. On the former, and in the northern half of the county, the hemlock, teasel, and nightshade are abundant, and here too we find the vervain (*Verbena officinalis*), the starwort (*Aster tripolium*), hound's tongue (*Cynoglossum officinale*), the crested cow-wheat (*Melampyrum cristatum*), and the frog-bit (*Hydrocharis morsus-ranae*). Roses, too, flourish better here than on the higher soils. Among plants that are lovers of the chalk are the campanulas—the hairbell, *C. rotundifolia*, and *C. glomerata*—the potentillas and scabious, hawkweeds (*Hieracium*), eyebright (*Euphrasia*), yellow-wort (*Chlora perfoliata*), the carline and musk thistles (*Carlina vulgaris* and *Carduus nutans*), the gentian (*G. amarella*), and many of the orchids, besides the rock-rose and the pasque-flower (*Anemone pulsatilla*). On the Greensand are the St John's-wort, the bird's-foot (*Ornithopus perpusillus*), bilberry, ling, foxglove, black alder (*Rhamnus Frangula*), tansy, wormwood, yellow toad-flax, and the lily of the valley. Among its characteristic weeds are the corn marigold, here known as "goolds"; the chamomile (*Pyrethrum parthenium*), called "mayweed"; the spurrey (*Spergula arvensis*), called "beggarweed"; the white goosefoot (*Chenopodium album*), called "father"; and the rest-harrow (*Ononis arvensis*) which generally goes by the name of "cammock."

There are fine oaks on the northern clay slope of Ampthill Park, some of which were doubtless too old for shipbuilding in Armada days. Many are still in full vigour, others mere barkless wrecks. The elm flourishes, and fine avenues of them are to be seen at Wrest Park. Luton Hoo, on the chalk of the south, is famous for its beeches, and from Caddington to Houghton Park on the northern edge of the sand the holly is commonly seen. Tall rows of the Lombardy poplar are conspicuous near Shefford, at Gravenhurst, Bromham, and elsewhere; and there is a famous cedar avenue at Southill Park. The box, occurring on Dunstable Downs, is believed to be a native of the county.

Bedfordshire cannot boast of a salmon river, though one or two of these fish have been recorded as casual

visitors caught in eel-traps. Trout have been introduced both into the Ivel and the Ouse, but in the latter
river they have been largely exterminated by pike, and
though they did well in the Ivel it was found that they
would not rise to the artificial fly. Barbel have been
established in the Ouse and have done well, and the
river is famed as one of the best for bream in England.

The county is not favourable either to reptilian or
batrachian life, and hence the fauna of these is poor.
The natterjack toad, though found just over the border
in Cambridgeshire and Herts, is not known to occur,
nor has the sand lizard been recorded, though found in
Huntingdonshire. The great water newt and the common newt are found, but not the palmated newt. The
slow-worm or blind-worm is common, as is the equally
harmless grass snake, especially in the Ouse meadows,
and the adder frequents the bracken and woods of the
Greensand and is tolerably abundant on the boggy
ground near Westoning. The green tree-frog has established itself in Woburn Park as an escape, and another
small exotic frog has acclimatised itself in the gardens
of Bedford, where it clicks monotonously through the
summer nights.

The draining of the fens on the north and the higher
farming and enclosures of the last century have not
been without their effect on the bird life of the county.
A hundred years ago the buzzard, hen-harrier, raven,
and kite, nested with us, but now even sparrow-hawks,
crows, and magpies are lessened in numbers, and the
birds just mentioned are extinct as breeding species,

and the hobby is a great rarity. Still, the ornithology of Bedfordshire has not been exhaustively studied, and

Young of Little Owl (*Athene noctua*)

it is possible that wider knowledge may lessen some of the rarities. On the chalk-hill range the Norfolk plover or stone curlew was nesting in the 'eighties, but unfortunately no longer does so, though snipe still breed with us, and the redshank is increasing as a breeding species. There are no heronries in the county, though herons are common.

If Bedfordshire has thus in some respects a limited avifauna, it is strong in others, for it lies on an important line of migration and consequently is visited by great numbers of birds of passage, some of them, such as the black redstart, the great grey shrike, and the waxwing, being rare. The crossbill is a regular winter visitant, and recently a certain number have nested. The tufted duck, formerly only a winter visitor, is now a well-established breeding species. The little owl (*Athene noctua*), introduced about 30 years ago into Northamptonshire, has spread widely and numerously over the county. The goldfinch and the starling appear to be increasing in numbers, especially the latter. The nightingale is fairly abundant.

The mammals call for no particular remark. The marten, polecat, and black rat have all disappeared and the badger is somewhat of a rarity. The otter, on the other hand, is very common on the Ouse and Ivel. The introduced grey squirrel has wholly supplanted our native species within the last few years at Woburn, and at least as far north as Tingrith and Steppingley.

8. Climate.

The climate of any country or district is dependent upon many factors, such as the latitude, the altitude, the direction and strength of the winds, the rainfall, the character of the soil, the nearness of the district to the sea, and all these factors are mutually interacting and interdependent.

Latitude has, naturally, a very great influence upon temperature, but there are many things which counteract and mitigate its effects, notably the proximity of the great oceans. These have a great effect in equalising temperature, and preventing extremes either of heat or cold. The lowest known temperatures occur in north-eastern Siberia, far from the Pole and in the middle of a great continent, in a latitude corresponding to not more than two or three hundred miles north of the Shetland Islands, where, owing to the warm ocean currents, the winters are often extremely mild. The greatest annual temperature range is found in the centres of continents and far from sea, and the most equable in sea-surrounded islands. Thus we have what are technically known as "continental" and "insular" climates; of which latter Great Britain is an excellent example, though it is abnormally mild owing to the prevalence of south-west winds which cause a movement of the warm surface waters of the Atlantic towards our shores.

The prevalent Atlantic winds are the main factors of our rainfall. Passing over the heated southern waters

Relief Map of Bedfordshire

(*From the raised map by Mr F. Hawkins Piercy*)

they become charged with moisture. The amount of such vapour that the atmosphere will carry depends upon temperature, and a lowering of the temperature causes proportionate condensation of the vapour carried, and its deposit as rain, snow, or hail. Temperature falls in proportion to the height of the elevation, and our south-westerly winds meeting with elevated land-tracts directly they reach our shores—the moorland of Devon and Cornwall, the Welsh mountains, or the fells of Cumberland and Westmorland—blow up the mountain slopes, become cooled, and at once deposit their vapour as rain. To how great an extent this occurs is well seen on referring to the accompanying map of the annual rainfall of England, where it will at once be noticed that the heaviest fall is in the west, and that it decreases with remarkable regularity until the least fall is reached on our eastern shores.

But Britain is surrounded with water, and it may be asked why winds other than those from the west and south should not contribute more to our rainfall. The reason is simple. The sea lying to the east of us is very small in area and shallow; while that to the north, though more extensive, is cold. The southerly and westerly winds therefore, sweeping over a vast extent of warm sea, can both receive and carry more vapour than those of the north and east, and in addition are, as already stated, more prevalent.

The above causes, then, are those mainly concerned in influencing the weather, but there are other and more local factors which often affect greatly the climate of a

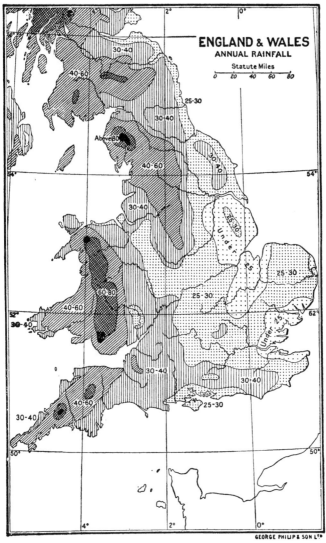

ENGLAND & WALES
ANNUAL RAINFALL

Statute Miles
0   20   40   60   80

30-40
40-60
25-30
30-40
Above 80
40-60
30-40
30-40
25-30
Under 25
40-60
60-80
30-40
25-30
Under 25
40-60
30-40
30-40
30-40
25-30
40-60
30-40

GEORGE PHILIP & SON Lᵀᴰ

(The figures give the approximate annual rainfall in inches.)

place, such for example as configuration, position, and soil. The shelter of a range of hills, a southern aspect, a sandy soil, will thus produce conditions which may differ greatly from those of a place—perhaps at no great distance—situated on a wind-swept northern slope with a cold clay soil.

Bedfordshire would be a very much drier county than it is if it depended for its rainfall upon the south-west winds. Irregular disturbances of the atmosphere, accompanied by electrical phenomena, are characteristic of the drier eastern counties, and occur chiefly between July and October. To these and the violent precipitation accompanying them we are indebted for a considerable part of our rainfall.

The average yearly rainfall for the whole county, calculated from the reports of a number of stations over a period of about thirty years, is approximately 24½ inches; the highest average fall being 33·22 in. in 1903, and the lowest 17·83 in. in 1870. The Kensworth rain-gauge, standing at the highest point (630 ft.) of those that furnish records, registered 42·11 in. in 1903—nearly 11 inches above any other year of the decade 1900–10. There are many places in the county from which records are sent to the British Rainfall Organisation in Camden Square; but it is to be regretted that from large areas, especially in the north, no records are received.

9. Population, Race Affinities, Type, Dialect, etc.

Bedfordshire has an area of 302,942 acres or about 473 square miles, and is the smallest county in England with the exception of Huntingdonshire, Middlesex, and Rutland. Its population in 1911 was 194,588. If we compare it with its immediate neighbours, it is nearly half as big again as Hunts and has more than three times the population; it has three-quarters the area of Hertfordshire, but less than three-quarters of its population; it has somewhat less than three-fifths the area of Bucks, but is only about one-eighth behind in its population; though not very much more than half as big as Cambridgeshire it has the same population within a fourteenth; and while less than half the size of Northamptonshire it has considerably more than half the population of that county. In other words Huntingdonshire has about 150 inhabitants for every square mile, Cambridgeshire 215, Buckinghamshire 262, Northamptonshire 336, Bedfordshire 364, Hertfordshire 409.

The history of the county makes it pretty certain that the original stock of its English inhabitants was Anglian in the north and east, and Saxon in the southwest. A general impression of this type suggests medium height, low tone in colour both of flesh and hair, and length and ruggedness of feature. Red or very dark hair, high-coloured cheeks, chubbiness, and round faces, seem comparatively uncommon. In accounts of the introduction of lace-making and straw-plaiting it has been asserted that Lorrainers settled at Luton and Huguenots at Cranfield, but as such assertions have been invariably made without production of any evidence it would be unsafe to build upon them. Undoubtedly a considerable fraction of the present population has immigrated from elsewhere; and it is probable that only a very small proportion of the present inhabitants of Bedford or Luton are descendants of Bedfordshire families. There has also been an influx from outside into Sandy and Kempston, and probably into the other more populated parts of the shire.

The testimony of language presents similar difficulties. The spread of uniform education, ease of communication, and the general use of reading, have destroyed local peculiarities and "standardised" the matter and manner of speech. The masters and mistresses of the village schools are seldom natives of the county, and have probably lost when they come here most of the tincture of their own country-sides.

In the Introduction to his *English Dialect Dictionary* Dr Wright classes Bedfordshire in the group of counties that extend from Rutland and Northamptonshire between the Thames and Wash to the coasts of East Anglia and Essex. This forms his "Third Division," and he subdivides it into five groups, the second of which he calls "Middle Eastern." This contains Mid-Northamptonshire, Bedfordshire, Huntingdonshire, and most of

Hertfordshire and Essex. From the linguistic point of view it is an interesting part of England, as it comes well within the area that has given English its modern form. As with other matters of compromise, two bodies of men uniting and speaking closely-allied dialects with marked differences are likely to suppress the differences and make the most of their common stock. It was upon this South Mercian and Anglian frontier that Anglian met Saxon, and three hundred years later the common dialect that this meeting had evolved was again affected by the presence of Danes speaking a kindred variety of the same group of languages. It was here then that all the more special and cumbrous prefixes and terminations were earliest discarded, that contraction took place, and that the phrase was moulded after an easier and simpler fashion. A circle of "twenty miles radius from the centre of Rutland is the cradle of the new English we now speak," says Kington Oliphant, and he adds "the land enclosed with a line drawn from the Humber through Doncaster, Derby, Ashby, Rugby, Northampton, Bedford, and Colchester, helped mightily in forming the new literature." A hundred years ago Mr Batchelor, a Lidlington farmer of sense and literary inclination, wrote an essay on the dialect of Bedfordshire, but few of the phrases he quotes do more than illustrate pronunciation. He points out that the greatest peculiarity in this respect is the complete absence of the sound *ou* as in *house*. It is represented by *ew*, the *e* being pronounced as in *ten* or *bent*. This is still markedly the case.

Batchelor left some record of Bedfordshire vocabulary in which the following words are noteworthy:—

Hurlock (kerlock) = hard seams of chalk. (The ploughman is careful not to plough enough on the chalk hills to disturb the noxious hurlock); Henmould = thin-stapled crumbling clays; Clung = tough and sticky (applied to clay); Livery = sad, heavy, and friable (clay); to lodge = be laid by rain (of crops); to gather = tiller = sprout out at base of stalk or root; Kid = wad = bean-pod; Bun = bean stubble; Kosh = pod; Maume cart = dung cart.

The following are, or were, mainly in use north of the Ouse:—

Seblip = sower's scuttle or basket for grain; Gotch = pitcher; Dol = small sheaf to mark tithe-shock; Trap = plank bridge; Whittawe = worker in white leather; Kimnel = kiver = flat wooden bowl for setting milk; Gudgel = hole or pool of muck or stagnant water; Fōw = scour or cleanse a ditch or pond; to binge = soak (a tub, etc.) in water; Quockend = choked; Garld = white spotted with red; Rafty = fusty, rusty (of bacon, etc.); Beeld = hovel; Gage = horse harness; to o'erwemble = overturn; Yelm = parcel of straw laid ready for thatcher; Yelmer = one who tends yelms; Tasker = thresher; Tranter = corn factor; Badger = licensed corn factor; Cottered = embarrassed; Avern = squalid, slatternly; Unked = strange, uncouth; Frem = vigorous, in good condition; Brokled = fragile, easily broken; Gain = apt, handy; Ungain = inapt, unhandy.

The following are in present use:—

Pightle = a close of ground varying from about a rood to a few acres; Twitchel = a narrow passage (e.g. near Shillington Church); Pendle = the upper layers of limestone, in which it lies in broken lumps.

10. Agriculture.

According to the statistics of the Board of Agriculture crops and grass cover some five-sixths of the area of the county, in the proportion of three to two. Grass shows a tendency to increase at the expense of the arable, and the same tendency may be seen in all the surrounding counties except Cambridgeshire, where arable is to grass as three to one, and still encroaches upon it.

In addition to the arable and grass land there are about 13,300 acres of woods, coppices, and plantations, and about 1200 acres of waste and heath that afford some rough grazing.

The average yield of the four chief corn crops, in bushels per acre, calculated upon the decade 1898–1907, is :—Wheat 30·97, Barley 30·59, Oats 40·92, Beans 29·10. If this be compared with the average yield of England, Bedfordshire wheat falls short by nearly 3 pecks, barley by 2¾ bushels, oats by more than a bushel, beans by about three pecks. If compared with the yield of a county of good soil and high farming, such as Lincolnshire, the wheat and barley are deficient by more than 3 bushels, the oats by 9, and the beans by 5. As compared with Huntingdonshire, however, the yield of Bedfordshire wheat is about three pecks greater than that of Hunts, that of barley 1¾ bushels better, and that of oats a bushel more, though it falls short of the Cambridgeshire yield by as much as 10 bushels. In beans however the yield is nearly 7 bushels above that of Huntingdonshire.

The chief changes that have taken place in Bedfordshire farming during the past century have been mainly due to four causes, (1) enclosure, (2) the growth of scientific agriculture, (3) improvement in means of communication, and (4) the competition of imported food-stuffs. Before the middle of the eighteenth century only a very few Bedfordshire parishes had been enclosed; as late indeed as 1790 three-quarters of the whole number were still farmed in open fields, that is to say, arable fields were divided into strips of $\frac{1}{4}$, $\frac{1}{2}$, 1, or (rarely) up to 4 acres, each in separate ownership. These were termed open or common fields. An owner of 25 acres might have his land in 50 strips scattered over the parish. Meadow land also was often held in strips. Under the open-field system farms were made up of intermingled patches. After harvest the grazing of the land was thrown for a time into common pasture, with a proportionate right of use, and occupiers of farms and cottages had rights or customary privileges of grazing and collecting fuel on waste, woodland, and warren. This system imposed upon each farmer such a rotation of crops and fallow as found favour with the majority. It was a bar to originality, experiment, and improvement. But enclosure, while it removed these drawbacks, and undoubtedly added greatly to the food-supply of the nation as a whole, was carried out in Bedfordshire, as in most parts of England except Lincolnshire and Norfolk, in such a way as to entail very serious disadvantages.

Small yeoman farmers and cottagers, in compensation

for their loss of common pasturage rights, were allotted parcels of ground that would not support the one or two cows they were accustomed to keep, but were attractive as additions to the farms of their wealthier neighbours. They were generally soon sold, and the money spent; milk passed out of the daily diet of agricultural labourers, and the small holder almost disappeared. Meanwhile the farmers who survived were in a far better position than their predecessors had been to take advantage of all that science could teach them. The main disadvantage under which they still suffered was the difficulty of obtaining leases, or leases free from inconvenient restrictions. This was however often mitigated in practice by the good relations existing between landlord and tenant. Such was certainly the case in Bedfordshire, under many of the larger landowners at least, and farms commonly remained for generations in the tenancy of the same families, though the strict bond was often but an annual one. The rapid increase of enclosure took place during the last ten years of the eighteenth century and the early years of the nineteenth, and thus coincided with a period in which much was done to establish agriculture on more scientific foundations. Francis, fifth Duke of Bedford, who came of age in 1786 and died prematurely in 1802, was as well known as Mr Coke of Norfolk among the pioneers of this movement. Experiments were carried on at Woburn to test different methods of ploughing, planting, and sowing, forms of plough and drill, breeds of sheep and cattle, the influence of soils and manures upon crops, the exact consumption and cost of live stock, and the qualities and culture of new roots and grasses. The "Woburn Sheepshearing" became as famous as the "Holkham Clipping," and both were attended by farmers, not only from many parts of England, but from the continent and the United States. At these "Shearings" ploughing competitions were held, and new methods

Ploughing on heavy land at Bletsoe

and machines were introduced and exhibited: they were indeed the precursors of the Agricultural Shows and Meetings of to-day. Bedford was one of the first places in which the manufacture of agricultural implements was established on a large scale, and Howards' ploughs and binders soon won a world-wide celebrity. The Bedfordshire Agricultural Society was founded as early as 1801, under the auspices of the Duke of Bedford and others, and it still flourishes, though it is a matter of regret that less than half the farmers in the county give it their support.

The interest which Francis, fifth Duke of Bedford, had shown in agriculture was shared by his brother John, who succeeded him, and experiments were continued at Woburn by him and by his successors. In 1876, some forty years later than the Rothamsted institution was started by Sir John Lawes, the Woburn Experimental Farm was established in Husborne Crawley, near Ridgmont Station, under the directorship of Dr Voelcker. Under his management, and that of his son who succeeded him, very valuable investigations have been, and are still, carried out, especially in connection with the chemistry of manures and the chemical and mechanical action of different soils. Grain has been grown for thirty successive years upon equal plots of similar soil, with and without manure, and with natural and artificial manures of various kinds; and the results have been carefully recorded and published. Exhaustive experiments have been made in the rotation of crops, and the manurial value of oil-cake and other foods for cattle and sheep. Attention has been given to the loss which farmyard manure suffers in making, storing, and moving, and to methods by which its virtues may be preserved. By pot-culture chemical experiments have been carried out which supplement the Mendelian researches of the Cambridge Laboratory in quest of a wheat which shall combine the milling excellence of Canadian, and the yield of English, wheats. An experimental fruit farm has also been established in the neighbourhood, at which important experiments have been made with regard to methods of tree-planting, the effect of a grassed surface upon the yield and quality of fruit-trees, and the nature and use of washes for the extermination of parasites. In 1896 the County Council established a County Agricultural Institute and Farm School, for which the Duke of Bedford supplied both the land and the building. A score of pupils were there taught the elements of farm work, ploughing, hedging and ditching, and then sent for a year of practical farming, after which they returned for a course of agricultural chemistry and geology. An unforeseen occurrence brought this promising scheme to an end, but its resuscitation would undoubtedly prove of benefit not merely to the county but to the country.

11. Market Gardening.

It is not easy to draw the line very exactly in Bedfordshire between agricultural and market-gardening occupation. Many vegetables which were considered to belong exclusively to garden culture at the beginning of the nineteenth century are now cultivated over an extent of hundreds of acres, and garden produce is grown on small allotments, and on some of the fields of large agricultural holdings. From the neighbourhood of Eaton Socon in the north-east to Henlow in the south, and extending eastward to Potton and westward from the Ivel to Willington, through Southill, Clophill, Maulden, Flitwick, and Greenfield, as far as Toddington, and by Clifton and Shefford to Upper Stondon, market-gardening steadily increases in area. The two kinds of soil which most favour it are the deep-stapled loams of the valleys of the Ivel and Flitt, and those gravels on the banks of the Ivel and Ouse that are overlain by a rich alluvial deposit. In some cases natural causes have mingled sand and clay, and produced a soil chemically nutritive and of a consistence and depth suitable to the penetration of deep-rooting vegetables. In other cases the soil is at once fertile and "quick" enough to supply early markets. Sandy and Biggleswade are two of the chief centres, and they send vegetables by the Great Northern Railway to the manufacturing towns of the north of England, and to London, to the amount of as much as fifty or even a hundred tons a day in the busy seasons. As well as the large trade beyond the limits of the county much garden-stuff is conveyed by train or cart to Luton and Bedford. Potatoes, carrots, onions, parsnips, parsley, celery, marrows, and many other vegetables cover many hundreds, indeed some thousands, of acres. This area is constantly extending, and at Greenfield raspberries, asparagus, and other garden produce may be seen growing within a yard or two of the shrinking area of Flitwick bog, upon which they steadily encroach. Flowers, too, appear at intervals among the vegetables; at Broom in Southill, for instance, many acres are covered by a regular succession of flowers for the London market.

The cultivation of woad, noted in old guides as characteristic of Bedfordshire, has long disappeared from this county.

The Bedfordshire yield of potatoes stands at 5·88 tons to the acre on the average of the ten years ending in 1907, and is slightly in excess of that of England as a whole and above the yield of all the surrounding counties except Hertfordshire. The yield of 9·58 tons in 1908 was exceptional, and is the highest quoted for any county in Great Britain in that year: it is only approached by Middlesex, Lancashire, and half-a-dozen Scotch counties. But this includes the Bedfordshire potato cultivation in all soils and under all conditions; there is no available return for the special soils of the Sandy neighbourhood. In the cultivation of carrots our county is only beaten in acreage by Cambridgeshire and Lincolnshire, owning 876 acres under that vegetable to the 1564 and 1393 of those counties; indeed it claims one-eleventh of the whole carrot acreage of Great Britain. In the cultivation of onions it takes the first place, as its 785 acres (mainly in the Biggleswade neighbourhood) amount to nearly a quarter of the whole onion area of Great Britain. Its other vegetables and flowers are grown upon the 3849 acres returned by the Board of Agriculture as under "other crops," and this area is only exceeded by Essex, Lincolnshire, Cambridgeshire, the West Riding of Yorkshire, Middlesex, Kent, and Surrey. In small fruit generally its acreage and yield are insignificant. Few counties return so small an acreage of orchards—1073 acres. The "prune damsons,"

upon a soil that is an eastern extension of the fertile vale
of Aylesbury, are a notable feature of Eaton Bray. The
fruit is large and of excellent quality, and is very profit-
able in good years. Kempston too possesses a large and
valuable walnut orchard. But Bedfordshire does not
take serious rank as a fruit-growing county.

12. Industries—Lace=Making. Straw=plaiting.

Lace-making, which played such an important part
in the economic history of Bedfordshire during the
eighteenth and nineteenth centuries, almost disappeared
from the county twenty years ago. It is true that the
number of lace-makers returned in the census of 1901
was 1148 out of a population of 171,707, or about one
in 150. But it must be remembered that when that
return was made almost all the 1148 were old or elderly
women, and very few of them were regularly employed.
For all practical purposes lace-making was dead in
Bedfordshire before the end of the nineteenth century.
At the beginning of it the growth of straw-plaiting had
to a great extent ousted lace-making from the south-
western part of the county, but elsewhere the great
majority of the women and girls above the age of seven
practised it. In the north, where the soil is poor and
farmers were unwilling to pay for unnecessary labour in
the winter, boys and even men are said to have taken to
lace-making for want of other employment. The lace-
schools were the only schools that girls attended in the

villages, on every week-day but Saturday, and it is stated on good authority that a girl who had learned lace-making for six years was, after the age of thirteen, no longer an expense to her family. Twenty years ago it would have been difficult to find a girl who was learning this art, and more difficult still to find a village lace-school.

About the beginning of the present century several ladies of this county and of Buckinghamshire began to interest themselves in the restoration of the industry. The matter was brought before the County Council in 1907, a small sum was voted, and classes were established in some twenty towns and villages. Many girls have been already taught, and the results are considered so far satisfactory that the Council has lately increased the grant in aid. It is too early to form any definite judgment as to whether there is any likelihood of a permanent and extended market for the lace, or whether girls can be taught on Saturdays alone to do what the girls of the past took every week-day but Saturday to learn. They worked too in the summer "from six or seven in the morning till sunset, and from eight or nine in the winter mornings till ten or eleven at night, by the light of a candle magnified by a glass globe, or upturned carafe, filled with water."

The circumstances of the introduction of lace-making into Bedfordshire and the neighbouring counties, or even into England, are still very obscure. The story which connects the industry with the residence of Catharine of Aragon at Ampthill is unsupported by evidence and improbable. There is little evidence that

Lace-maker at work

(Bobbin wheel, iron candlestick, and glass globe for focussing light, on the table)

the art was practised in Europe at all before the sixteenth
century. Fuller is probably near the mark when he
says, in his account of lace-making in Devonshire,
"Modern the use thereof, not exceeding the reign of
Queen Elizabeth"; and adds, "it saveth some thousands
of pounds yearly, formerly sent over seas, to fetch lace
from Flanders"—a reference consistent with the connection which students of lace have pointed out between
that of the South Midlands and the productions of
French and Belgian Flanders. Although he says
nothing of "bone lace" under Beds or Northants, in his
account of Bucks he speaks of "much thereof made about
Owldney (Olney) in this county." Shakespeare speaks
of "maids that weave their threads with bone," and the
art was probably already established in Buckinghamshire
at least, for we find that shire complaining, as early as
1623, that the "bone-lace trade was much decayed,"—
clear evidence that the trade was not then a new one.
Besides its familiar names of "thread lace," "bone
lace" (from the bone bobbins) and "pillow lace," it has
been called "English Lille" lace. Buckinghamshire
has generally been regarded as the principal seat of the
industry, and it has been characteristic of the west
rather than the east of Bedfordshire, although it spread
at one time over much of the county. In the eighteenth
century the neighbourhood of Olney, Stony Stratford,
and Newport Pagnell formed the centre of the industry
in Bucks, the neighbourhood of Towcester in Northants,
and in Bedfordshire the district between the Bucks
border and Woburn and Bedford.

The early history of the straw industry in Bedfordshire is as obscure as that of lace-making. The accounts
which ascribe its introduction to James I, and the
Napiers of Luton Hoo are not supported by any direct
evidence. But this industry differs from that of lace-
making in one respect. The construction of a tissue
from straws, grasses, or any other materials that could
be easily interwoven, goes back to the dawn of civilisation, and light hats were doubtless thus made in many
places, but the question of interest is not the introduction
of the art, but its specialisation and concentration in
North Hertfordshire and South Bedfordshire. On the
11th of August 1667 Pepys was visiting Hatfield in Hertfordshire, and he tells us that "the women had pleasure
in putting on some straw hats which are much worn in
this country, and did become them mightily." Just
before that date Fuller wrote in his *Worthies*: "When
Hartfordshire wheat and barley carries the credit in
London, thereby much is meant (though miscalled),
which is immediately bought in, and brought out of, Hartfordshire, but originally growing in Bedfordshire about
Dunstable and elsewhere." Fifty years later, in September, 1724, we find Heaton "soliciting a patent to make
hats of bent or straw which would have been extremely
prejudicial to thousands of poor people about Hempstead
in Herts, and Luton and Dunstable in Beds. At the
same towns £200 have been turned on a market day in
straw hats only, which manufacture had then thrived
in those parts above one hundred years, and children as
well as grown people maintained themselves by plaiting

wheat straw and working it for hats and other uses."
A protest was backed by the influence of the Duke of
Bridgewater, who then owned what is now Lord Brown-
low's seat at Ashridge, and the patent was refused.
Dalton's *Traveller* is not dated, but internal evidence
shows that its Bedfordshire ·material was collected not
later than 1776. It tells us that "Dunstable carries on
two large manufactures, one of straw hats and the other
of lace," and calls Luton "a handsome town, [with] a
considerable manufactory of straw hats." Pennant
writing in 1780 says nothing of the trade of Luton,
which he describes as "a small dirty town," but tells us
that at Dunstable "a small neat manufacture of straw
hats and baskets and toys maintains many of the poor."
At the beginning of the nineteenth century Lysons says,
"The straw manufacture prevails, and has of late
much increased, in the neighbourhood of Dunstable and
Toddington and on the border of Hertfordshire. The
employment is not necessarily so sedentary as lace-
making, for the straw may be plaited by persons standing
or walking. The earnings even of those who make the
coarse plait are higher than those of the lace-makers,
and the profit of making the fine plait is very considerable." About the same time (1808) Batchelor tells us
that "straw-plaiting, which was formerly confined to
the chalky part of the county, has spread rapidly over
the whole southern district as far as Woburn, Ampthill,
and Shefford," and while arguing that the earnings
quoted are exaggerated by ignoring the time occupied
in sorting and bleaching and the expense of the straw,

he admits that straw-plaiting "has on the average been productive of more advantage to the poor than lace-making." An interesting extract from Arthur Young's diary of September, 1801, shows how highly he appreciated the advantages of the industry. "At Dunstaple: went to meet a person who instructs people in plaiting straw, and I bargained with him at 30*s.* a week for a

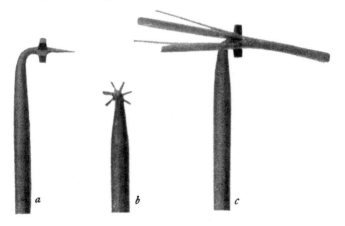

Straw-splitting Implement
(*a. side view, b. front view, c. in action*)

[Norfolk] girl to be instructed: a month will do: that is £6, and the journey there and back about £4. 10*s.* For £10 I shall be able to introduce this most excellent fabric among our poor. The children begin at four years old, and by six earn 2*s.* or 3*s.* a week; by seven 1*s.* a day; and at eight and nine 10*s.* and 12*s.* a week. This will be of immense use to them." He should have

added that such wages could only last during certain
limited periods of the year.

By the beginning of the nineteenth century, then,
the trade was well established.   But it was restricted to
the use of whole-straw plait, and could not compete with
the finer work of Tuscany, which was largely imported
in the last quarter of the eighteenth century.   The
method of dividing the straw with a knife was clumsy
and productive of irregularity in the material.   The
invention of the straw-splitter was inevitable and gave
a new impetus to the work, but it is uncertain when
and how it first came into use.   One story attributes
the invention to the French prisoners at Yaxley Barracks
near Stilton, who are said to have sold their plait in
Luton;   another to a youth at Chalfont St Peter's in
South Bucks.   The splitter that passed into common
use consists of a stalk of iron or brass about three inches
long and one-eighth of an inch in diameter.   The lower
end may be held in the hand or inserted upright in a
block of wood.   The upper end tapers, and is bent at a
right angle as a sharp spur of half or three-quarters of
an inch, around which is set a star of small blades,
generally varying from four to seven in number, beyond
the centre of which the sharp thin spur extends for
about half an inch.   This spur is thrust into the hollow
of the straw, which is pushed towards the blades, and
becomes evenly divided into a corresponding number of
strips.   The split straw is then flattened by being
passed through a small rolling mill of two contiguous
cylinders of box or other hard wood, and such may still

be seen upon the wall in many a Bedfordshire cottage, though few are now in use. The straws are carefully selected, and those of particular districts, Barton for instance, enjoy a special repute. They are cut between the last knot and the ear and are about nine inches

Rolling-mill for flattening the Straw Plait

long. Their selection, cutting, and distribution, was for a long time a considerable industry in itself.

Towards the end of the first quarter of the nineteenth century fashion turned once more to the finer straw-plait of Italy, and the more energetic of the Bedfordshire hat-makers entered upon the task of competition. Conspicuous among them were the Wallers of Luton, to whom the Bedfordshire trade indirectly owes its present prosperity. Experiments were made with different bents and various methods of preparing wheat-straw, and an important advance took place when the plaits were first joined by sewing instead of knitting. Tuscan straw was imported, and plaited here. White chip, of willow wood, had already been introduced at Dunstable as early as 1795, and in 1820 finer chip was made, woven in a loom, and sewn into hats and bonnets. Some of it was made of black poplar; some, of the kindred Lombardy poplar, was imported from France; and various kinds of rice straw were also employed. By the middle of the century many improved forms of the "whole" or "Dunstable" straw were made; and a variety of fancy forms were produced by waving, fluting, and other methods. Horsehair and other materials were already in use. St Albans had become an important rival, and more expensive work was imported from Italy and Switzerland. The turn-over at Luton, where there were thirty principal manufacturers, and branch factories of London houses employing 200–300 workmen apiece, was about £1,500,000 per annum. The extension of the Midland Railway to London in 1868 gave Luton an advantage of which it rapidly availed itself, and since that time the straw trade has dwindled into insignificance at Dunstable. A very few years later the first cargo of Chinese plait was shipped to England, and soon China was supplying much of the plait used. In 1890 the Japanese turned their attention to the Luton market, and at the present time the bulk of the plait comes from them. They have already copied and beaten the Italians, and threaten to rival the fancy plaits of Switzerland, and the braid and "crinoline" of Germany. The amount of straw plait now made in Bedfordshire is small, not one per cent. of the whole, and is chiefly for the American market. But the villages around Luton have not suffered by the change. Far more money is earned by the work of hat-making, trimming, and straw-dyeing, than was earned of old by plaiting. In several centres of the neighbourhood, even as far north as Ampthill, there are factories in which a number of workers are employed, and many are brought to them from the surrounding villages by a regular system of carriage. In Luton, the population of which has increased in a century from 3000 to 40,000, large dye-works have been established, and almost the whole trade of straw-dyeing has been secured. To the making of hats has been added their decoration, and thousands are sent out daily to many parts of the world. Nor is the industry now restricted to straw even in its widest application. Chiffon, velvet, and chenille have already been added; and a trade in beaver hats is projected, if not established. This development of millinery has further led to a large industry in the manufacture of cardboard hat-boxes and cases. So small a plant is requisite that by the side of the large firms many small establishments manage to thrive. Few towns, indeed, show more certain signs of a prosperity which is not only growing but widely diffused.

Luton

At Bedford, besides Howards' old-established "Britannia" works, are the engineering works of W. H. Allen, which cover much ground west of the railway and have created a large new suburb; also the "Pyghtle" works for garden seats and other ornamental woodwork; the Grafton foundries, and other factories of motor parts and machinery, on the Ampthill and Luton roads. At Luton are made pumps, boilers, marine and hydraulic engines, motors, ball-bearings, stoves and ranges, tool-making machines, aeroplanes, etc.

The foregoing are the most noteworthy industries of the county but there are a few of lesser importance which deserve mention. Such are those supplied by the abundant osiers and rushes of the Ouse and Ivel. Osiers are cut and peeled at several places, notably at Biggleswade, in the neighbourhood of Bedford, and at Pavenham; and a good deal of basket-work is done at all three. The rush industry of Pavenham has been traced back to the end of the seventeenth century; and early in the eighteenth this place appears to have had a repute for rush-matting beyond the borders of the county. Although now only local, the industry continues to thrive, and in many articles the rush and osier are combined in basket-making. The rush used is the bulrush of botanists, a sedge (*Scirpus lacustris*), not the reed mace (*Typha*), with its club-like cylinder of bloom, to which the name of bulrush is popularly given.

At Odell there is a small fell industry, at Harrold a manufactory of leather, and at Potton one of parchment and chamois-leather.

13. Minerals.

The mineral wealth of Bedfordshire is only such as can be obtained from limestone and chalk, clay, sand, and gravel. Beds of gravel are distributed widely over the county; they are best developed along the older beds of the rivers, but occur at intervals where they have been left by the drainage of glaciers. They contain fragments of rocks of many kinds, sandstones, limestones, chalk, and flints of local origin, mingled with others brought by glaciers from the far north and north-west. When the gravel has been dug it is screened, or sifted; the larger stones being used for rough building material, and the smaller, combined with sand, for our roads and paths. The pits at Biddenham and Kempston will serve to illustrate this industry. Many of the churches of the county, especially in the north and south-east, where no better building material was available, have been largely built of pebbles and small boulders.

Sand is quarried at Sandy, Flitwick, Leighton Buzzard, and other places; it is used for mortar, bricks, metal-moulds, filter-beds, and various other purposes. The white sands of Heath, near Leighton Buzzard, for example, are employed in glass-making in Birmingham. About 100,000 tons of gravel and sand are quarried yearly. In some places, as at Cainhoe, Sandy, and elsewhere, the sand exists in indurated beds, the particles being cemented by the infiltration of peroxide of iron. Such sandstone (called ironstone when it contains an

Sand-pit at Sandy

excess of iron) has been used to build many of the
churches of the central part of the county. A few other
buildings are constructed of it, those of Biggleswade
waterworks for instance; but it would be difficult to find
any houses or cottages of sandstone, except one or two
at Sandy. In former days local limestone was largely
used for building in the district between Stagsden,
Wymington, Yelden, and Bedford. The churches, and
many of the older houses and cottages, farm-buildings,
and walls, are constructed of it in many villages within
that area. But it is not a sufficiently good building-
stone to send out of the county, and few of the quarries
are now open. It has been generally superseded by
brick.

Limestone is also burned for lime. Till very recently
there were lime-kilns at Biddenham Ford End, and at
Clapham; and Lime Street in Bedford preserves the
memory of a lime-kiln which is marked on a seventeenth
century plan of the town. Chalk is a very pure form of
limestone, and the lower beds, or Chalk Marl, consist of
calcareous clay passing gradually, as they rise, into
purer chalk. They are quarried at Sundon, in several
places along the face of the chalk escarpment under
Streatley and Sharpenhoe, at Dunstable, Arlesey, and
Totternhoe. The chalk is burned for lime, which is used
to make mortar or cement. The harder kinds of chalk,
such as the Totternhoe stone and Melburne rock, and the
Hurlock, or chalk rock, have been much used of old for
building. They are often called clunch, and have been
employed with flint, boulders, sandstone, and pebbles in

the fabric of most of the churches in the south of the county. Totternhoe stone was for a long time in considerable repute, and is said to have been used not only for the west front of Dunstable Priory church, but in the interior of Westminster Abbey. The quarry

Totternhoe Clunch Quarries
(*Norman earthworks at the summit*)

is named among the gifts which Henry I bestowed on Dunstable Priory at its foundation (about 1130). The stone is not now used for building.

Flint appears in the walls and towers of many churches in the south of the county. Sometimes it is the chief material; elsewhere, as at Luton and Barton, it is used rather as an ornament to form a chequer with clunch or other stone. It is also used in road-making.

The clays of the county provide material for brick-making. Wherever the clay occurs, large brickfields may be found near the towns, and smaller amongst the villages. The ruins at Someries are the oldest brick buildings in the county (fifteenth century), and the bricks may have been imported. The brick remains of Warden Abbey (sixteenth century) and of Houghton House (1615) may be of local manufacture, but are more likely to have come from elsewhere. The more important brickworks at which bricks are made to be exported from the county are conspicuous by their groups of tall chimneys, those upon the Oxford clay for instance south of Bedford, and the brick and tile works on the gault clay at Arlesey. On descending from the north into the gault plain, in the neighbourhood of Gravenhurst, it will be noticed that the more recent buildings are all of a compact light yellow or almost white brick. This light colour is due to a large percentage of lime in the gault clay, which passes up, indeed, into Chalk Marl. The variation from light buff to deep red seen in the bricks made elsewhere depends upon the amount of iron the clays contain. At Caddington and Luton are made the blue-grey " Luton bricks," the wet clay being dusted with powdered flint before baking.

The fuller's earth of Woburn had a great reputation as early as the seventeenth century. It is a soapy clay capable of absorbing grease, and was dug till recently in the neighbourhood of Woburn, but little is being extracted at present.

The curious nodules or lumps of stone known as the coprolites of the Greensand and the gault have been dug in great quantities at Potton, Ampthill, Henlow, Stanbridge, and Billington: it is said that 30,000 tons were extracted between 1870 and 1880. The "red" or Greensand coprolites, which owe their colour to iron, appear to have been formed by chemical accretion about animal or vegetable remains, which must have lain in water highly charged with phosphorus. They are found near the base of the Greensand, and often contain fossils from the Kimmeridge clay, which was denuded before the Greensand was laid down. The "black" coprolites of the gault contain a larger percentage of phosphates. The coprolites are ground to a powder, and treated with sulphuric acid to release the carbonic acid gas, leaving a phosphate of lime. This is used as manure to restore phosphoric acid and lime to exhausted soils. At the present time the industry has ceased, but it may be resumed whenever a method is devised by which the ground coprolite can be effectively applied to the soil without the additional expense of treatment with sulphuric acid.

14. History.

The British of Bedfordshire at the time of the Roman invasion (43 A.D.) were probably of the tribe of Catuvellauni, and were ruled by one of the princes of the family of Cunobelin (Cymbeline) from his stronghold at Verulam, which we now call St Albans. We know nothing of the Roman history of this part of Britain later than those stirring days when its inhabitants saw the rapid march of Suetonius Paulinus (60 A.D.) as he hastened back from Anglesey in a vain attempt to save London and Colchester from the fury of the revolted Iceni under Boadicea. Durobrivae (on the site of Dunstable) is the only Roman town recorded as within its borders, but villas and farmsteads doubtless grew up throughout the shire.

Saxon Sword found in Russell Park, Bedford

About a century and a half after the final withdrawal of the Romans the men of Wessex invaded the country of the Britons north of the Thames. In 571, says the Chronicle, "Cuthwulf fought the Britons at Bedanforda (Beada's Ford) and took four towns," including Lygeanbyrig, which was probably Limbury near Luton[1]. This conquest gave the West Saxons a claim to the country between the Ouse as far as Bedford, and the Lea. The growth of Mercia barred them from taking advantage of their claim for the next two centuries. Mercia extended its way southwards to London, and its greatest

[1] Lygea is the Lea, and Lygeanbyrig may be any place on the Lea— Limbury, or Leagrave, or Luton (Lygeatun).

king, Offa, was buried at Bedford, on the north bank of the Ouse, in 796. The invasion of the Norsemen soon confined the energies of the Kings of Wessex to the defence of the country south of the Thames, and when Alfred forced Guthrum to a treaty in 885 the boundary line between Wessex and the Danelagh was drawn "upon the Thames, then upon the Lea as far as its source (i.e. at Limbury), then straight to Bedford, then upon the Ouse as far as the Watling Street"; leaving Bedford and the greater part of the county to the Danes, while the south-west remained with Wessex. Thus, from the earliest recorded times, Bedford and the head-waters of the Lea were connected.

When Edward the Elder harried the Danish territory between the Ouse and Bury St Edmunds in 905 the east of Bedfordshire must have been laid waste. In 917 Luton gained distinction by beating a Danish force that came from Northamptonshire, and in the next year, while Edward was busy fortifying Buckingham, the Jarl Thurkytel and the chief men of Bedford made submission to him. He proceeded to Bedford and remained there for a month, strengthening the town, and building a *burh* or fortification on the south side of the river. A fine example of his work that remains is the King's Ditch, which encloses the part of the town that he added to the south of the Ouse. In the year 921 the Danes abandoned Huntingdon and established themselves at Tempsford, at the junction of the Ouse and Ivel. From that base they crossed the river and attacked Bedford, but the garrison sallied out and beat off their assault. It is possible, and even probable, that the earthworks at Tempsford, the small Danish fortified post with boat docks that lies on the south bank of the Ouse at Willington, the fortified position at Castle Mills,

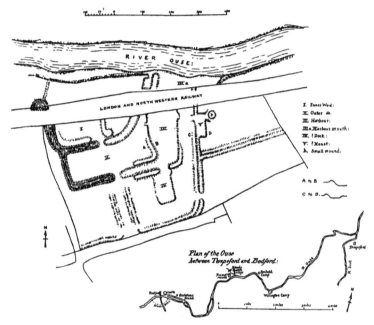

Plan of Willington Camp

and probably too the entrenchments at Howbury, are all connected with this event. The occupation of Bedford was part of a strategic movement admirably carried out by Edward the Elder and his sister, the Lady of Mercia, by

which they gradually advanced a chain of fortresses from
Staffordshire to Essex and drove the Danes steadily
before them. But, even when the central and south
Midlands were recovered, the Danes remained so thickly
settled in Northants and the counties to the north of it,
that the district of the Five Boroughs was accorded
special conditions of government. This drove a wedge
into Mercia and led to the incorporation of Bedfordshire
and the neighbouring counties with East Anglia.

A division of this part of England into shires took
place in the tenth century, and Bedfordshire assumed
approximately its present form. It was now in the
revived, but altered, diocese of Dorchester, the see (or
chief place) of which was removed to Lincoln in the next
century. Ten years before the end of the tenth century
the Danish attacks recommenced. The treacherous
massacre of Danes on St Brice's day in 1002 led to the
invasion of East Anglia by Sweyn in 1004. He conquered that district and sent his armies north and
south. They burnt Cambridge and made their way
through Buckinghamshire "down the Ouse till they came
to Bedford, and so forth as far as Tempsford, and ever
burned as they went, and so went back to their ships
with their booty."

William the Conqueror's army probably laid waste
part of Bedfordshire as he made his circuitous advance
upon London. As no favour was shown to the landowners
of the shire, it is probable that most of them took an
active part in Harold's defence of the country. With
the exception of Waltheof who married the Conqueror's

niece Judith, a few almsmen and officials who held
small estates, and some of the monasteries, every land-
holder was deprived but Albert the Lorrainer. Afterwards William used the forced labour of the men of
Bedfordshire and the neighbouring counties to build his
castle at Ely.

Henry I founded the town of Dunstable towards the
end of his reign. A few years earlier he had established
the Augustinian Priory there, and built himself a royal
residence, Kingsbury, which stood not far to the north
of the church. For several centuries Dunstable Priory
was often visited by our kings. Hugh de Beauchamp
had obtained the largest grant of lands in Bedfordshire
when Domesday Book was compiled, and he was almost
certainly Sheriff of the county at that time. At what
date a stone castle was built at Bedford is uncertain,
but no doubt the mound was at least strongly stockaded
and already in his keeping. When built, the castle was
of great strength and was held for some weeks by Miles
de Beauchamp against Stephen's army. Henry (afterwards Henry II) attacked the castle in 1153 and did
much damage to St Paul's church in the course of his
siege, but whether he took it is not known.

In 1166 the burgesses of Bedford obtained a charter
of incorporation from the Crown. For this they paid
forty marks and an annual rent to the Crown. At
the same time the county gaol was built at Bedford,
probably at the corner of Silver Street and High Street,
but possibly within the castle.

In the troubles that arose towards the end of John's

reign William de Beauchamp, the son of Simon, was one of the Barons who leagued together to demand a Charter of Liberties and on their repulse at Northampton came to Bedford, where he entertained them at his castle. John signed Magna Charta in June 1215, but when the Pope declared it invalid, the King summoned foreign mercenaries, who marched north under Fulk de Bréauté

Bedford Castle
(*As reconstructed by Mr Charles H. Ashdown, F.R.G.S.*)

and other leaders, spent a night at Dunstable, and went on to Northampton, ravaging the estates of the discontented barons on the way. In December Fulk arrived at Bedford and captured the castle, which the King bestowed upon him. William de Beauchamp was one of the barons excommunicated by name, and Fulk got his lands with the castle, and also the castles of Hanslope, Oxford, and Northampton. He soon became the tyrant and terror of the neighbourhood. He persecuted the monks of Warden, carrying off thirty of them prisoners to his castle at Bedford. In 1223 he resisted all demands for the surrender of his castles to castellans appointed by the Crown, and entered into relations with the disaffected Earl of Chester and Llewelyn of Wales. The crisis came in 1224. In the summer of that year actions were brought before the judges at Dunstable by several freeholders of the Manor of Luton demanding the restitution of lands of which Fulk had dispossessed them. The decision was given against him in every case, and his brother, William de Bréauté, acting by his orders, waylaid one of the judges, Henry de Braybrook, an important landowner in the shire, and imprisoned him at Bedford. The news of this outrage found the King and his chief ministers in council at Northampton in June 1224. All other business was put aside, and within a week the King arrived at Bedford, with the Archbishop, Hubert de Burgh, and an armed force. Fulk was away in the West treating with Llewelyn, but his brother William had dismantled the churches of St Paul and St Cuthbert, and had made every preparation to hold the castle. This stood upon a site which may be easily traced to this day. Its ditch ran along the High Street on the west, south of Mill Lane on the north, and down Newnham Road to the river on the east. The keep or main tower was on the mound in the south-east of that area on which there is now a bowling-green and a summer-house. North of the keep was the inner, and west of it the outer, bailey.

The ditches were almost certainly filled with water diverted from the river. After desperate fighting and a siege of nearly two months, the besieged held only the strong inner tower or keep. The walls of this were now undermined and temporarily underpinned with shorings of wood. On the 14th August, 1224, all was ready, the wooden supports were set on fire, and as the smoke rose into the interior through the fissures in the cracking masonry, the besieged saw that further resistance was hopeless. They hoisted the royal ensign, set free the imprisoned judge and other captives, and sent out with them all the women who were within the walls. On the next day the garrison were brought before the King. Three were pardoned on condition they went on Crusade, the chaplain was handed over to Archbishop Langton for punishment, and more than eighty, including William de Bréauté, were hanged upon the spot. Fulk came to Bedford and made his submission; he was deprived of all his possessions, banished the kingdom, and died abroad a few years later. The castle was dismantled, the walls "lowered," the keep destroyed, the ditches filled up and levelled, while the buildings of the inner bailey were stripped of their fortifications and given back to William de Beauchamp for a dwelling-house. Thus Bedford castle disappeared in 1224, and the Beauchamps did not long outlive it. Fulk de Bréauté had also built a castle at Luton, which was dismantled at the same time.

For the next forty years there is little to record in Bedfordshire but a couple of tournaments at Dunstable.

In 1263, when the quarrel between the Barons and the King led at last to actual warfare, Simon de Patteshall of Bletsoe, Hugh Gobion of Higham, Ralph Pyrot of Harlington, and Baldwin Wake of Stevington, all followed de Montfort. In February 1265, the Earl of Gloucester was to lead one side in a great tournament at Dunstable, but the meeting was forbidden by the King's Counsellors. Gloucester refused obedience, but was obliged to submit on the arrival of Simon de Montfort with a strong force. Their friendship had long been cooling, and this exercise of authority led to an open rupture. Gloucester left Dunstable in dudgeon, and he and de Montfort next met on the battlefield at Evesham. Amongst those who fell there with Earl Simon was John de Beauchamp, youngest of the sons of the William to whom the castle site had been restored. He was but just of age, and had "raised his banner for the first time" at the battle in which he fell, the last male representative of the Beauchamps of Bedford, on 9th August, 1265.

The year 1295 was the first in which Bedfordshire and the Borough of Bedford both began to send members regularly to Parliament. Only one other town in the county was ever represented—Dunstable—which made no return to the writ issued to it for 1311, but was represented in 1312 by two burgesses. This, however, was the only Parliament to which it sent members.

In the troubled reign of Edward II Dunstable was the scene of an assemblage of the discontented barons. After the king's death, Mortimer and the Queen held a festival at Bedford in 1328, and the next year the Earl of Lancaster encamped near the town at the head of an army, and was with difficulty induced to make submission without a fight.

In the reign of Henry VI, the weakness of the central government, and the lawlessness of the great landowners and nobles, were as noticeable in this county as elsewhere, and in 1439 large gatherings of gentlemen of the county and yeomen came into collision at Bedford, Lord Fanhope leading on one side, and Sir Thomas Wauton on the other, upon a dispute between rival Justices of the Peace. Lying between Northampton and St Albans, Bedfordshire must have suffered much from the disturbance of the Wars of the Roses, but no actual conflict took place within its borders.

In the sixteenth century Bedford Grammar School was founded. Much property in the county came into the hands of the Crown, by the dissolution and forfeiture of the monasteries, and by the alienation of Ampthill and other estates to Henry VII. Henry VIII constituted the several royal estates into one as the Honor of Ampthill, and frequently spent some weeks at Ampthill Castle to hunt in the neighbouring parks. James I also repeatedly visited the county for hunting, and stayed at Ampthill, or as a guest at Bletsoe, Toddington, or Houghton.

When the Civil War broke out the county as a whole declared for the Parliament. Many of the gentry of the county undoubtedly fought, and fought well, for the Crown, but "the King had not in Bedfordshire," says

Clarendon, "any visible party or one fixed quarter." Rupert successfully attacked Bedford and held it for a short time, and Dunstable was plundered in the midst of its Michaelmas fair. Newport Pagnell was largely garrisoned and pecuniarily supported by Bedfordshire throughout the war, and in June 1644, when Essex had been enticed into Dorsetshire, and Waller was vainly endeavouring to come up with the King, the west side of Bedfordshire suffered much from the Royal army. Between the 22nd and 26th of June parties of the King's large force of cavalry raided Dunstable, Leighton, Hockliffe, and Woburn, and threatened Bedford. When Brown advanced from Hertford to co-operate with Waller, the King drew off, and Bedfordshire was left in peace. After Naseby (June 1645) the King made a circuitous march from Cardiff to Grantham, and reached Huntingdon 24th August, but finding he was pursued and had not sufficient force to do anything serious, he hastened across Bedfordshire to Oxford. After plundering Huntingdon, his troopers stripped the northern half of our shire of everything on which they could lay hands. On 25th August part of this force was at Bedford, and Charles slept that night and the next at Woburn Abbey. This was the last visit the Royalist troops paid the county. In August 1646, the garrison of Newport Pagnell and the small garrison of Bedford, which consisted of eighty foot and forty dragoons (mounted infantry), were disbanded.

But though the county was now free from all menace of assault, it was to see the King once more within its borders. When the controversy arose between the Army and Parliament upon the question of the settlement of the kingdom, the army headquarters were moved from Reading to Bedford, 22nd July, 1647, and the King was brought from Caversham to Woburn. Cromwell and Ireton were constantly at Bedford and kept up the negotiations that they had begun at Reading. Charles was treated with all respect, the Earl of Bedford, who had been for some time in retirement at Woburn, attended him, the Earl of Cleveland visited him from Toddington, as did many other gentlemen from the neighbourhood, and he was allowed the services of secretaries and chaplains. It was at Woburn that the Army's Proposals, drawn up by Ireton "the wise penman," were laid before him. The Army leaders were opposed to the Presbyterian element, which was strong in London and the Commons House, and wished to make their own arrangements with the King without the interference of Parliament. The King thought that he could play the Independents and the Presbyterians one against the other, and so gain time while he negotiated for the help of the Scots. When Ireton and other Army officers had their final interview with him at Woburn Abbey the King, who had been encouraged by private assurances that the Parliament and the Presbyterians in London would support him against the Army, surprised even his own counsellors by the "tart and bitter" words with which he received them. He insisted upon the legal establishment of the Church, and told them boldly that without him they could effect no settlement. Sir John Berkeley, who was in attendance upon him and had already advised him that the terms offered by Ireton were more favourable than he could have hoped to get, ventured to whisper a warning, and the King made some attempt to be more conciliatory. But his words had had their effect. Colonel Rainsborough stole away from the Conference and galloped over to Bedford, where he harangued the troops and inflamed them against the King. Berkeley hurried after him, but was too late to counteract the impression which the report of the King's answer had already made; and when news suddenly arrived of the riot in London, the invasion of the Houses of Parliament, and the flight of the Speaker and many Members, the fervour was fanned to flame. The Army left Bedford on the 29th of July to march upon the City, and two days later the King was removed from Woburn. He never had another chance of accepting such reasonable terms.

During the centuries that have followed the Civil War few events of more than local concern have taken place within the county, and the part played in public affairs by Bedfordshire men is indicated in the section on the Worthies of the County.

15. Antiquities.

We have no written record of the history of our land antecedent to the Roman invasion in 55 B.C., but we know that Man inhabited it for ages before this date.

The art of writing being then unknown, the people of those days could leave us no account of their lives and occupations, and hence we term these times the Prehistoric period. But other things besides books can tell a story, and there has survived from their time a vast quantity of objects (which are daily being revealed by the plough of the farmer or the spade of the antiquary), such as the weapons and domestic implements they used, the huts and tombs and monuments they built, and the bones of the animals they lived on, which enable us to get a fairly accurate idea of the life of those days.

So infinitely remote are the times in which the earliest forerunners of our race flourished, that scientists have not ventured to fix either the date or the length of the periods into which they have arranged them It must therefore be understood that these divisions or Ages—of which we are now going to speak—have been adopted for convenience sake rather than with any aim at accuracy.

The periods have been named from the material of which the weapons and implements were at that time fashioned—the Palaeolithic or Old Stone Age; the Neolithic or Later Stone Age; the Bronze Age; and the Iron Age. But just as we find stone axes in use at the present day among savage tribes in remote islands, so it must be remembered that weapons of one material were often in use in the next Age, or possibly even in a later one: that the Ages, in short, overlapped.

Let us now examine these periods more closely. First, the Palaeolithic or Old Stone Age. Man was now in his most primitive condition. He probably did not till the land or cultivate any kind of plant or keep any domestic animals. He lived on wild plants and roots and such wild animals as he could kill, the reindeer being then abundant in this country. He was largely a cave-dweller and probably used skins exclusively for clothing. He erected no monuments to his dead and built no huts. He could, however, shape flint implements with very great

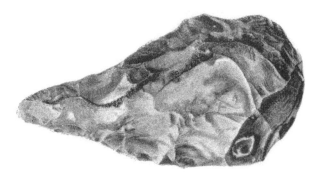

Palaeolithic flint Implement from Biddenham

dexterity, though he had not as yet learnt either to grind or to polish them. There is still some difference of opinion among authorities, but most agree that, though this may not have been the case in other countries, there was in our own land a vast gap of time between the people of this and the succeeding period. Palaeolithic man, who inhabited either scantily or not at all the parts north of England and made his chief home in the more southern districts, disappeared altogether from the country, which was later re-peopled by Neolithic man.

Neolithic man was in every way in a much more advanced state of civilisation than his precursor. He tilled the land, bred stock, wove garments, built huts, made rude pottery, and erected remarkable monuments. He had, nevertheless, not yet discovered the use of the metals, and his implements and weapons were still made of stone or bone, though the former were often beautifully shaped and polished.

Between the Later Stone Age and the Bronze Age there was no gap, the one merging imperceptibly into the other. The discovery of the method of smelting the ores of copper and tin, and of mixing them, was doubtless a slow affair, and it must have been long before bronze weapons supplanted those of stone, for lack of inter-communication at that time presented enormous difficulties to the spread of knowledge. Bronze Age man, in addition to fashioning beautiful weapons and implements, made good pottery, and buried his dead in circular barrows.

In due course of time man learnt how to smelt the ores of iron, and the Age of Bronze passed slowly into the Iron Age, which brings us into the period of written history, for the Romans found the inhabitants of Britain using implements of iron.

We may now pause for a moment to consider who these people were who inhabited our land in these far-off ages. Of Palaeolithic man we can say nothing. His successors, the people of the Later Stone Age, are believed to have been largely of Iberian stock; people, that is, from south-western Europe, who brought with them their knowledge of such primitive arts and crafts as were then discovered. How long they remained in undisturbed possession of our land we do not know, but they were

Bronze Age Palstaves found at Wymington and Silsoe

later conquered or driven westward by a very different race of Celtic origin—the Goidels or Gaels, a tall, light-haired people, workers in bronze, whose descendants and language are to be found to-day in many parts of Scotland, Ireland, and the Isle of Man. Another Celtic

people poured into the country about the fourth century
B.C.—the Brythons or Britons, who in turn dispossessed
the Gaels, at all events so far as England and Wales are
concerned. The Brythons were the first users of iron in
our country.

The Romans, who first reached our shores in B.C. 55,
held the land till about A.D. 410; but in spite of the
length of their domination they do not seem to have left
much mark on the people. After their departure,
treading close on their heels, came the Jutes, Saxons,
and Angles. But with these and with the incursions of
the Danes and Irish we have left the uncertain region of
the Prehistoric Age for the surer ground of History.

In the middle of the last century a doctor at Abbeville,
M. Boucher de Perthes, succeeded at last in persuading
the scientific world that what we now term Palaeolithic
flint instruments were neither freaks of nature nor
shaped by accident, but of human workmanship.
Among those who visited Abbeville, and studied the
sites at which M. Boucher de Perthes had made his
discoveries, was the late Mr James Wyatt of Bedford.
He saw that the gravels of the Ouse presented conditions
similar to those of Abbeville and Amiens; and he had
for a long time studied the gravel-pits of Biddenham.
They are sunk in a cap of gravel at some sixty feet above
the present level of the river, and lie about half a mile
north of its modern bed, and within the southern sweep
with which the river curves from Bromham Bridge to
Bedford. At a depth of about fourteen feet the gravel
rests upon a mass of cornbrash, the highest member of

the oolite limestone. Mr Wyatt had already collected
from near the base of the gravel, and from similar
gravel beds in the neighbourhood, fossil bones of extinct
mammals such as the cave-bear and hyena, extinct
forms of rhinoceros, hippopotamus, deer, oxen, and
elephants; and a careful examination of the gravel had
proved that it was of fresh-water formation, and full of
fragments from the drift clay. It was clear too that the
whole mass over which the road runs from Bedford to
Bromham Bridge had been an island of limestone about
and upon which the Ouse had left deposits of clay and
gravel, which it brought down from the north-west when
it was cutting its way as a mighty river through the bed
of boulder clay that spread from the Clapham hills to
those of Kempston and Stagsden. Thus well prepared,
Mr Wyatt patiently watched the gravel-pits, and at last
his opportunity came. In the April of 1861 a fresh
excavation was carried to the base, and eagerly scanning
every spadeful, it was not long before he espied two
excellent examples of types already familiar at Abbeville.
The news of his discovery was heralded by English
geologists as marking an epoch, not merely because it
proved the presence of Palaeolithic man in Britain, but
because the discovery carried our knowledge a step
farther than it had been advanced by the French finds.
"One step at least we gain," says Lyell, "by the Bedford
sections, which those of Amiens and Abbeville had not
enabled us to make. They teach us that the fabricators
of the antique tools and the extinct mammalia were
co-eval, or in other words posterior to the grand sub-

mergence of central England beneath the waters of the glacial sea."

Nearly thirty years later, in 1890, an even more remarkable find was made by Mr Worthington Smith in the extreme south of the county, upon ground very unlike the valleys of the Somme or the Ouse. This discovery again was the result of no mere chance; it came, as Mr Wyatt's had come, from observation, reasoning, and patient care. Mr Worthington Smith had found worn ochre-stained Palaeolithic flints on the surface of the Kensworth hills and the Houghton fields, and had endeavoured for eight years to account for their presence there. He has told the whole story, and it is a very interesting one, in his *Man the Primeval Savage*, which everyone in Bedfordshire should read. On one of his many visits to the Caddington brick-pits he noticed a thin band of flint in the face of one pit, and found that it consisted of sharp flakes. When this was uncovered it proved to be an old land-surface on which lay flakes, cores from which flakes had been struck, flint instruments, and a number of still unbroken flints so piled as to make it evident that they had been placed there by human agency, and still bearing traces on them of the different soil from which they had been extracted. All these were buried under a bed of mud, formed partly of boulder clay, partly of the tertiary clays that are found in patches on the heights of Caddington and Kensworth. The sections of this clay showed clearly that it had been brought down in a liquid stream from higher ground. This Palaeolithic "floor" lies at a height of 520 feet above sea level and some 150 feet above the average level at which water now lies in the chalk. So gradually did the mud cover up the floor that Mr Worthington Smith has been able to reconstruct flints from the flakes, finding in some cases as many as a dozen which fitted one another accurately. The edges are as sharp as those of flakes struck off to-day, and show no sign of being rolled by water. All the flint instruments found on this floor are Palaeolithic, but none are of the spearhead type that occurs at Biddenham. They are generally scrapers, oval blades, narrow knife-like blades, punch-shaped, or hammers. They show no sign of ochre-staining, but vary from black through indigo to grey and white, and have a china-like surface. In the clay with which they are covered were found specimens of the ochre-stained worn kind, which it was clear had preceded in time those which they helped to bury. An admirably-arranged case illustrates the discovery, and begins the series of flint instruments exhibited at the British Museum.

When Neolithic man arrived Britain was an island. The clay hills of Bedfordshire were covered with forest growth, the plains were marshes, and the new-comers would at first inhabit the chalk uplands and the gravel ridges by the rivers. They were a dolichocephalic, or long-headed race, and made long barrows or tumuli in which they buried their dead. Their stone axes and chisel-like weapons called *celts* have been found at Pavenham and Felmersham, Kempston, Biddenham, Cardington, and along the Ivel valley, and more generally in the neighbourhood of Dunstable. There too are still to be seen some remains of the long tumuli, though many have been destroyed by the plough.

Relics of Bronze Age man also occur in Bedfordshire. Their hut-hollows and round tumuli may be seen on the Dunstable downs, and their socketed bronze celts have been found at Wymington, Toddington, and near Dunstable, and along the Ouse and Ivel valleys. Two of their camps, Maiden Bower and Waulud's Bank, still remain; and the encampments at Totternhoe, Mossbury Hill near Bedford, Caesar's Camp at Sandy, and John of Gaunt's at Sutton may have been originally occupied by them. Maiden Bower is a circular camp of slightly irregular outline, which measures 775 by 750 feet and stands about a mile west of Dunstable. It covers more than ten acres and is pierced by five entrances. The rampart is strongest on the south-west, where it measures 28 feet through at the base and is 10 ft. 6 ins. in height. The surrounding ditch originally measured about 20 feet in width, but its outer edge is now scarcely visible. From the number of flint implements found in this camp it is probable that it was first thrown up by Neolithic Man and later occupied by people of the Bronze Age. Waulud's Bank lies just north of Leagrave station. Its shape would be a segment of about half a circle, and it covers about ten acres.

A long time before the arrival of the Romans the Celts had taken to using iron for their cutting tools and weapons. Their coins, uninscribed and inscribed, have been found near Dunstable and in the Ivel valley:

those that can be dated are of the age of Julius Caesar and bear names of British kings, whose capital was at Verulam (St Albans).

The Roman remains are not of great importance, though they include specimens of many different objects —coins, pottery, glass, instruments, ornaments, and utensils in bronze, brass, iron, and silver. No kind of inscription has been found in the county except oculists' stamps and a seat pass, and but one piece of sculpture. But sufficient evidence exists of the presence of Roman or Romano-British settlements at Bedford and thence up the Ouse in the direction of Irchester (Northants), east to Sandy, south of Sandy through the Ivel valley, and south-west through Shefford along the southern edge of the Sandy district to Woburn, and at Dunstable and its neighbourhood in the south.

Fragments of pottery and coins of the Roman period have been found at Biddenham, Elstow, Cople, Cranfield, Northill and Caldecote, Old Warden, Haynes, Maulden, Flitton, Toddington, Higham Gobion, Shillington, Clifton, Stotfold, Caddington, Luton, and Kensworth. Mr W. Ransom, of Hitchin, has some fine bowls from Arlesey and Astwick, many of them Samian, and some of them of the Castor and Upchurch ware, specimens of which may be dated as of the first and second centuries. A bronze steelyard weight was found at Bromham. A well still exists in a field at Biddenham in which fragments of pottery, bones of animals, and a piece of Roman sculpture in relief were found half a century ago. Another well, 120 feet in depth, was opened close to Maiden Bower, containing Roman fragments; and in the immediate neighbourhood have been found pottery, coins, and other objects, and a paste intaglio portrait of Carausius. In Bedford remains have occurred in Horne Lane, Castle Lane, the High Street, and Potter Street (or Cardington Road). A gold coin of Vespasian was found at Tottern-hoe, a store of about 1000 (dating 196–270 A.D.) at Luton Hoo, and others at Yelden and Willington and elsewhere.

At Stanford, north-west of Clifton, two rooms described as "vaults" were discovered eighty years ago.

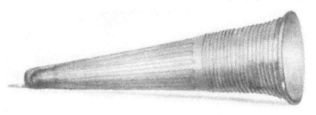

Anglo-Saxon glass Drinking Cup found at Kempston
(*10½ inches long, width of mouth 3½ inches*)

Each was 15 feet by 12 feet and 5 feet in depth. One contained bronze pans and jugs, cooking utensils, Samian and other pottery, remains of six amphorae or wine-jars, a flute, and a bone-handled knife. In the other were found silver tweezers, silver buckles, glass vases of different colours, beads, Samian pottery and wine-jars. At Shefford a Roman cemetery was unearthed in 1826, and amid numerous articles of interest were a handsome flat bronze bowl with a ram's-head spout and ornamental handle, vessels of glass, a leaden eagle, and an amphora.

Samian ware, and coins dating from 14–356 A.D. were also found at Shefford. At Sandy (which is not now identified with Salinae) the principal finds have been at Chesterfield, between Galley Hill and Caesar's Camp. Besides coins dating from 28 B.C. to 450 A.D., pottery, including a rivetted Samian bowl and Castor ware, and glass, there have been found a silver ring with a cornelian, a bronze plaque with a head of Mercury in relief, bronze bowls, and a sword.

We may now turn to Anglo-Saxon remains. To the west of Bedford on the Clapham Road, between Bedford and Newnham, and at Toddington and elsewhere, several graves have been opened. But of greater interest and importance is the discovery at the Kempston gravel-pits in 1863 of a burial ground of some extent which appears to have been long in use. Here both burnt and unburnt bodies were found, and the opinion formed by those who studied this burial ground at the time of its discovery was that the ashes belonged in many cases to an earlier use of the cemetery, that the intrusion of unburnt bodies had often displaced and disturbed earlier urn-burials, but that at a later time the two modes were practised contemporaneously. As cremation is considered to be characteristic of Anglian, and unburnt burial of West Saxon graves, the burial ground at Kempston gives testimony, consistent with such other evidence as we have, of the mingling of the two kindred nationalities. Moreover, the brooches discovered here are of two types—round, considered to be West Saxon, and cruciform, East Saxon or Anglian.

Several of the Kempston urns are adorned with incised patterns. One, found in the same neighbourhood, is remarkable for a small disc of greenish glass inserted in the bottom before the clay was fixed; its shoulders are adorned with incised patterns, and the lower half is deeply fluted. A large urn from Sandy, 9 inches high and 3 feet in girth, has elliptical bulges at intervals of its circumference.

We have in Bedfordshire some good typical earthworks. The small rectangular camp at Tempsford, ramparted and deeply moated, 120 feet by 84, must be the "work" which the Danes threw up there in 921. The "King's ditch" at Bedford bounds the Saxon *burh* which Edward the Elder erected south of the Ouse. A large series of "motte-and-baileys," otherwise known as mound-and-court castles, most historians now think were of Norman origin, being the rough and quickly-made strongholds which alien conquerors would require; instances are to be found at Bedford, where the castle was held for the king by succeeding members of the Beauchamp family and where stone defences were later added to the earthworks; at Cainhoe, near Clophill, the head of the Daubeny barony; at Odell, Thurleigh, Poddington, Ridgmont, and Totternhoe, all connected with the barony of Odell; at Eaton Socon, Yelden, Chalgrave, Toddington, Flitwick, Sutton, Tilsworth, and Meppershall, where either some sub-tenant of William I's or some rebel of Stephen's reign, imitating his more powerful neighbours, erected a small personal stronghold.

Cainhoe Castle Hill

(Stronghold of the Daubeny family from 11th to 14th century)

16. Architecture—(*a*) **Ecclesiastical.**

Before the Norman style of architecture was introduced into England in the middle of the eleventh century, churches were built of wood and stone, and their style is still called Saxon. We have record of a stone church taking the place of a wooden one at Studham just before the Conquest. The characteristic of Saxon architecture is the so-called *long and short work*, an alternation of vertical and horizontal pieces of stone forming the *quoins* or corners of the building, and serving both as ornament and frame for the rubble walls, into which the horizontal pieces bond them. Other features are triangular instead of semi-circular heads of windows and doorways, and balusters dividing window-openings, though these latter continued in use to a later period, as in the tower of St Mary's in Bedford. The tower of Clapham church, except its top storey which belongs to the twelfth century, is generally recognised as of pre-Conquest date. The tower of St Peter's in Bedford has been much tampered with and rebuilt outside, but inside there is a good specimen of the triangular-headed window. The lower part of the tower at Stevington has a Saxon doorway, and round-headed windows in the north and south fronts, which now open into aisles. The windows are double-splayed, and one of them still has in it remains of the pierced slab of wood which once filled the central aperture.

The Romanesque or Norman period lasted for about

Norman Doorway, St Thomas's Chapel, Meppershall

a century after the Conquest. It is characterised by massiveness, semi-circular arches, flat shallow vertical buttresses, and small window-openings. The capitals of the columns and pillars, and the ornaments of the arches of important doorways, are often grotesque, sometimes reproducing motives of earlier northern ornaments, and often consisting of simple zigzags or rows of billets. Important doorways are recessed, and show a succession of arches beneath the main one, each in a plane behind the plane of the one next above it; and the jambs of these arches are often filled by small corner columns. The west front of Dunstable is a good specimen of a late Norman recessed doorway, the ornament of which is partly classical, partly northern. Sometimes the doorway has a carved relief in the tympanum or space between the lintel and the arch-head above, as the north door at Elstow, or the south door of the tower at Thurleigh. The towers of St Mary's in Bedford, Kempston, Meppershall, and Cranfield show Norman work; Norman doorways may be seen at Little Barford, St Peter's in Bedford, Cranfield, St Thomas's chapel in Meppershall, and two interesting specimens at Kensworth. The most notable interior Norman architecture is in the nave of Dunstable, the eastern part of the nave at Elstow including the clerestory windows, and the nave at Everton; and several churches, such as Kempston, Knotting, and Poddington, retain their Norman chancel arches.

After the middle of the twelfth century the style grew lighter and more elegant, and gradually passed by

Nave of Priory Church, Dunstable

a period of transition to that which is called Early English. This period may be roughly dated from 1190 to 1290, and is marked by the introduction of the pointed

South door, Eaton Bray Church

arch. The transition from Norman to Early English is well illustrated by the west front of Dunstable with its big recessed Norman doorway flanked on the north by an Early English doorway, from which it is separated by a piece of Norman arcading of interlacing arches under a pointed arch of which the south arc has Norman ornament, the north Early English moulding. Above this arch and the doorway north of it runs an excellent specimen of Early English arcading, a series of recessed pointed arches supported by grouped columns. The interior of Elstow church again illustrates the transition, the

N. Transept   S. Transept

St Mary's Church, Felmersham
(*Showing stages of lancet development*)

eastern part being pure Norman, the western transitional with one arch at least thoroughly Early English. Grace, proportion, delicacy, restraint in ornament, and the beauty of contrasting light and shadow, are the notes of this style when it arrives at its perfection. Elaborate mouldings with deep undercutting produce marvellous effects of light and shade, and the slender columns, or groups of columns, whose lines are continued above in the

Nave arcading, Eaton Bray Church

separate mouldings, have a grace and elegance that is absent from the heavier Norman. The two churches which are best worth careful study are those of Felmersham and Eaton Bray. The former of these has a central tower, transepts, chancel, and nave with aisles. Its original east window is gone, the two central lancets of its west window have been at some early date combined into one central light, and the roof has been cut down, but with these exceptions the church stands almost exactly as it was built in the first half of the thirteenth century. The mouldings of the arches under the tower and in the nave are good specimens of Early English, which may also be studied at Harrold, Pertenhall, Chalgrave, Studham, and other churches, but they are surpassed by the masterly work in the north arcade and aisle at Eaton Bray. In this church the capitals of the columns also show the conventional Early English foliage at its best.

As the thirteenth century drew to a close the Early English style developed gradually into the Decorated, which is sometimes called Edwardian, because it extended over the reigns of the first three Edwards. The crosses erected to mark the halting-places upon the route by which the body of Queen Eleanor was brought to Westminster, and her tomb in the Abbey, show the excellence that sculpture and architectural ornament had attained by the end of the thirteenth century. Two of these crosses were erected in Bedfordshire, at Woburn and Dunstable, but were destroyed by the ruthless ignorance of the Parliamentarian armies in the

Civil War of the seventeenth century. The beautiful wrought-iron screen on the north side of Queen Eleanor's tomb at Westminster was the work of a Bedfordshire iron-worker, Thomas of Leighton Buzzard (1294). There is good iron-work on the church doors at Eaton Bray, Turvey, and Leighton.

The chief features of the transition from Early English to Decorated are the abandonment of lancet arches for arches around an equilateral triangle, the development of tracery in windows, various changes in the section of mouldings, less depth in undercutting, the introduction of crockets and highly-ornamental finials, the use of the ogee arch, and of triangular gables over trefoil openings in canopies of niches. The tooth ornament of Early English disappears and is often replaced by the ball flower[1], which is characteristic of Decorated in many places, but not common in Beds. The most notable feature of this period, however, is the development of window tracery. In the Early English style two or more separate lancet lights were placed side by side under a common arch. Where there were three the middle one was sometimes the tallest, as in the east end at Meppershall. Where there were two of equal height a blank space or tympanum was left between the heads of the lights and the arch above, and this was often pierced with a circular or quatrefoil opening, as in the transept at Felmersham. In all these cases the lights were independent and only in juxta-

---

[1] The ball flower is a spherical outer case about a ball, the case gaping with a trefoil opening in front and showing the ball.

Elstow Church

(The choir and central tower were wrecked)

position, separated by some width of walling. Gradually this space was reduced to the moulded casing of the several lights, but so that the framework of each light was independent of the other, and thus "tracery" came into being. A beautiful example, marking the entrance of the Decorated period, may be seen in the east window of the north aisle of Sundon. In some cases as many as five lancets with trefoiled heads were ranged together under one rear-arch and the whole space above filled with a network of stone mouldings inclosing quatrefoil or lozenge-shaped lights representing the meshes of a net. This is very characteristic of the early Decorated period, and examples may be found at Langford, Sutton, Dunton (east window), Wymington, Sundon, and elsewhere. Gradually the lines of the tracery grew to be an organic whole, as in the east window of the south aisle at Dunton, and in windows at Yelden and Chellington. Clifton (nave and chancel), Sundon, Yelden, Tempsford, Higham Gobion, Shillington, Wymington, St Paul's Bedford, Westoning, Northill, Salford, and Leighton Buzzard are all worth study for different features of this style. Westoning is notable for the effect of grandeur given by the height of the nave arcades. At Salford the wooden porch is an interesting feature.

In the later development of Decorated two features prepare us for a change. Many of the smaller window-openings are square-headed, as for instance at Wymington, although the tracery is essentially Decorated. Here and there too we find windows in which two of

St Lawrence's, Wymington

the mullions or vertical divisions between the lower
lights are carried up straight through the tracery to the
arch-head above; the east window at Chellington is an
example. But, in the main, there was no transition
period, and the architecture of the next period, the
Perpendicular, was a new departure, which developed
after the Black Death had checked all building for a
time, and was both widespread and rapidly adopted.
It is characterised, as its name implies, by the perpendicular (and horizontal) arrangement of the tracery, by
flattened arches and rectangular panelling, by elaborate
vault-traceries, especially fan-vaulting, by increased
window space, flattened roofs, and towers without
spires. Most of the churches of Bedfordshire were added
to or altered under the influence of this style. The
high-pitched roofs—the original position of which can be
often traced where they abutted on the east side of the
western towers—-were cut down and surrounded with
battlements, towers were rebuilt, porches added, and
new windows inserted, especially in the aisles and
clerestories. The Perpendicular period lasted from the
latter part of the fourteenth century till the dissolution
of the monasteries (1530–1540), though there are some
later instances, such as the north nave arcade of Campton.
The chancel screens and the remains of the rood lofts
that are to be found in several Bedford churches date
from this period; e.g. at Pertenhall, Oakley, Felmersham,
and Eaton Socon. Dean has a magnificent wooden roof
adorned with angels carved in wood; Barton one with
figures of eagles. Eaton Socon, Colmworth, Flitton,

Stevington Church

(Showing " long and short work" in the lower part of the tower ; windows of Decorated period)

Tingrith, Marston Moreteyne, Cople, and Willington are interesting Perpendicular churches. The most perfect piece of work characteristic of the period is perhaps at Luton, in the arch which opens from the north of the chancel to the Someries chapel. It gives an excellent example of Perpendicular panelling.

Most of the churches north of the Ouse have spires; but south of the river there are none except at Leighton Buzzard, Chellington, Eyworth, and Ridgmont which is modern. The spire of Leighton Buzzard is a fine specimen of Early English work. It is of the "broach" type, capping the tower directly, and not rising from within a parapet. Those of Pertenhall and Souldrop are also broach spires. The Perpendicular spires rise from within the socket of a parapet and seldom combine so well with the square tower as do the broach spires, unless the junction is adorned by well-proportioned pinnacles. Colmworth and St Paul's Bedford are instances of the parapet type. In Norman and Early English churches the tower very often stood at the intersection of chancel, nave, and transepts. St Peter's Bedford, St Mary's Bedford, Leighton Buzzard, Toddington, Felmersham, and Meppershall are all of this plan. Most of the towers of Bedfordshire churches are at the west end, but there are some exceptions. Langford tower is south of the south aisle. Stevington and Sundon have towers at the west end, but enclosed between the aisles. At Elstow and Marston Moreteyne the towers are isolated. The early Clapham tower was probably part of a stronghold. It is altogether out of

All Saints' Church, Leighton Buzzard

proportion to the size of the church, which was originally a mere chapel of Oakley.   The manor house at Clapham actually abutted upon the west side of the tower, and the church was probably at one time a manorial chapel. The tower of Felmersham is the one good Early English tower.   Odell is an excellent specimen of a Perpendicular tower, but perhaps the most elegant specimen of this style in the county is the tower of Cockayne Hatley. The carved woodwork at Warden is excellent, and few churches possess such wood-carving as that of the stalls at Cockayne Hatley.   But in neither case is the work English: it has been brought in recent times from the Netherlands.

The churches of the oolite district are generally built of local limestone.   Poddington, Farndish, and some other churches of the north-west use Northamptonshire limestone and sandstone.   Those in and near the central sandy district are wholly or in part of sandstone.   Chalk rock, especially Totternhoe stone, and flint are used in the south, and in Luton clunch and dressed flints are used chequer-wise.   In the clay districts pebbles are largely employed; and pebbles, chalk rocks, flint and sandstone are combined according to convenience of locality.   In the case of larger churches which have been built by the more wealthy monasteries or landowners limestone has been used which is not of local origin.

17. Architecture—(*b*) Domestic.

The cottages of Bedford are described in the beginning of the nineteenth century as generally consisting of two, three, or four rooms with a small piece of garden ground,

Cottages at Ampthill

frequently not exceeding a tenth of an acre. In the north-east of the county they were built with "daub and wattle," or, as it is locally termed, "stud and mud." Bricks and tiles were in more common use in the southern part of the county, and limestone was the ordinary material used in the oolite district of the north-west, in the triangle between Stagsden, Yelden, and Bedford. The chief change that has taken place in materials is the more widespread use of bricks. The older cottages of the limestone district are of stone, but those built within the last fifty years are usually of brick, and brick is in general use in the north-east as well as the centre and south.

The most noticeable feature of the Bedfordshire cottages is the semi-circular or segmental brow with which the roof often undulates over the upper windows. It apparently originates in thatching, but nevertheless is often retained when tiles replace thatch. Other features are the cream-coloured tint with which the plaster is washed, and the framework of wood in the earlier brick buildings.

The dwelling-house of more importance than a mere cottage or hut, but not aspiring to the dignity of a fortress or palace, was for many centuries after the Norman Conquest of a simple type. The main part of it was the hall, an oblong room with an unceiled roof. To right and left of this there were added one or two private rooms for the convenience of the family, and rooms for storage and treatment of food and drink— the buttery and kitchen. The only upper rooms, or *solars*, were in these wings. Matthew Paris, the chronicler of St Albans, speaks of Panlin Peyvre, who died in 1251, as "an incomparable builder of manor houses," and tells us that at Toddington he built a hall, a chapel, and private rooms, with other buildings, all of stone with lead roofs, and planted orchards and constructed fish-ponds in the grounds. Some twenty years later the Cainhoe property was divided between three heiresses, and the chief residence is described as Cainhoe Hall, with porch, chamber, and cellar towards

Cottages at Cardington

the east, bakehouse, dovecot, garden, fish-ponds, and three barns, and another chamber of limestone to the west.

The common type of old-fashioned manor house has preserved the tradition of this disposition. The central oblong mass of the house represents the old hall, though it may now consist of three storeys and be divided into rooms and passages; the two gabled wings at right angles to it have grown out of the additional rooms. It was probably not till the sixteenth century that it became common to insert a floor upon beams over the lower part of the hall, so making a ground-floor room, or perhaps two, and a passage to connect the two ends of the house. If the height was sufficient the roof space was also divided into rooms, and dormer windows were inserted in the roof. The old moated manor farm-house at Marston Moreteyne still preserves considerable remains of the timber work which formed its small central hall centuries ago. The central part of the house now consists of a ground floor, an upper storey, and a roof space above: but on the first storey the upper parts of the curving beams that formed some of the arched supports of the old hall roof have been kept where they were wanted to help in constructing partition walls for bedrooms and landings.

Harlington manor house till lately had the date 1396 upon it, and an old view of it given in Dr Brown's *Life of Bunyan* shows the relative size and importance of the central block, though internally it had already been divided into three storeys. Many farm houses in Bedfordshire were once manor houses. Campton manor house represents a late sixteenth-century development of the central hall with gabled wings. It is interesting too as still retaining the screens which were erected across the entrance, but within the hall, to give privacy and avoid the draught caused by the constant opening and shutting of the hall door during the service of dinner. The Campton screens are late Elizabethan woodwork. One of the sitting-rooms also is panelled with Elizabethan panelling and bears a bullet mark of 1645 upon it. Haynes Grange farm house has one very old half-timbered wing, and still retains the **H** shape given by a central

Cardington Manor House

hall connecting two wings. Cardington manor house farm is only about half of the original house, one wing and part of the centre. It dates from the sixteenth century and has a fine twisted chimney, several handsome chimney-pieces in the bedrooms, and panelled ceilings. A house in Cotton End, Cardington, the remains of one of similar type, retains an elaborately-moulded seven-teenth-century ceiling with coats of arms, human figures, and emblems in relief. Cardington and Marston manors are enclosed within moats still filled with water. Twisted chimneys of the same type as in Cardington manor house are a feature also of the old manor house of

Ruins of Houghton Conquest House

Mavorns, Bolnhurst, and of the brick building which is all that is left of Warden Abbey.

Houghton House has been dismantled for more than a century, but the general disposition of the rooms is clear. It was built between 1615 and 1620 for the Countess of Pembroke by Italians, if we may trust John Aubrey, who gathered his information thirty or forty years later. Although it does not retain the

Elstow Green, showing Moot Hall and remains of Cross

ordinary manor-house type, but approximates to a
square with corner towers, the old idea still dominates
the disposition of the apartments. Upon the ground
floor at least the eastern part is the service side, the
western contains the living-rooms; the whole of the central
part is practically the hall duplicated, a north hall and
a south hall back to back. The rooms are of a good
height throughout, and freely supplied with fireplaces,
and the upper floor extends over the halls as elsewhere.

Some houses have a very different origin, and are
adaptations of conventual buildings. Lord St John's
house at Melchbourne contains, with much alteration,
part of the buildings of the Knights Hospitallers.
Chicksands not only takes its shape from the part of the
Priory which it has preserved, but still shows much of
the old work. It belonged to the Gilbertines, an English
Order, and the shape and peculiarities of the house may
be all traced to its original structure and requirements.
Bushmead Priory still possesses its old refectory, annexed
to but not incorporated in the modern house. The
open roof appears to preserve much of its original timber
and form, dating apparently from the beginning of the
thirteenth century. At a later period it has been
divided into two storeys by the insertion of flooring.
The old house at Woburn, which had been formed from
the Abbey buildings, was pulled down just before the
middle of the eighteenth century; the chapel and
cloisters fifty years later. Of the house which rose
on the ruins of Newenham Priory nothing remains but a

few fragments of brick wall; and the "Britannia" works stand on the site of Caldwell Priory. The refectory of the Franciscan house at Bedford was removed about twenty years ago, and its place taken by a school playground and garden in Priory Street.

All that remains of Warden Abbey is a block of red brick with twisted chimneys that probably formed part of the Abbot's lodgings. It dates from the early sixteenth century. Nothing is left of Dunstable Priory but some interesting remains of the hospitium or guesthouse now forming part of a building in the Watling Street. The old "Moot Hall" which stands upon Elstow Green is a good example of a brick and timber building of the fifteenth or sixteenth century, and is thought to have served as the guest-house of the abbey.

In some cases castles or fortresses were converted into country dwelling-houses, but no complete example occurs in Bedfordshire, and most of the sites that were once fortified can now only be traced by earthworks. But one fragment is of great interest—the gate, gatehouse, and chapel of Someries, near Luton. It is a very early specimen of brick building, and was left unfinished at the death of Lord Wenlock in 1471. Its most interesting features are the adaptation of the old defensive structure to mere ornament. Over the main gateway, and over the postern that flanks it, are what appear at first sight to be pierced projecting galleries for attacking intruders from above the entrances; but they are not pierced, and serve only to give ornament and variety to the front. A farm house close to Bletsoe church still keeps the name of Bletsoe Castle, and is in fact a portion of one side of the large square of which that castle consisted. Odell Castle is now a country house, but it still contains considerable portions of the buildings that have succeeded one another on that site since the Conquest. Ampthill Castle, which was built in the first half of the

Someries Castle, Luton

fifteenth century, disappeared in the early half of the seventeenth, and Bedford Castle was dismantled as far back as 1224–5.

The mansions at Ampthill, Aspley, and Ickwell Bury were built at the end of the seventeenth century. The front of Hinwick Hall is a characteristic specimen of the semi-classical style in vogue in the middle of the eighteenth century. Luton Hoo has suffered so much by fire and rebuilding that it has lost most of its interest, and Wrest Park, which stands a couple of hundred yards north of the site of the ancient house shown in Kip's engraving, is a modern house in the style of a French château.

Old cottages, Luton

Of public buildings for civil purposes there are none of much antiquity. There are still remains, probably of the fourteenth or fifteenth century, of the Old George Inn at Bedford, and there are many old barns, such as the large tithe barn at Felmersham. Dove-houses of sixteenth and seventeenth century date are not uncommon, as for instance at Harrowden and Willington; and that at Ickwell Bury still retains its revolving ladder to visit the niches on each side successively.

18. Communications—Past and Present.

In judging of the probable direction followed by roads in early days it must be remembered that low-lying land was liable to flood, and was in many places little better than marsh. Thus the more important lines of communication were in pre-Roman days along the drier heights. In British times they were mere tracks. The most notable of these led along the northern edge and slope of the chalk from Norfolk through Cambridgeshire, Hertfordshire, South Bedfordshire, Bucks, and Oxfordshire to the Thames near Goring and Wallingford, and then across the Berkshire and Wiltshire downs to Bath or to Avebury. This is the famous Icknield Way. It was a broad track, not a road in the modern sense, but the Romans used it, and found it practicable without treatment. It entered the county at Dray's Ditches, passed Waulud's Bank near Leagrave, and ran through Dunstable south of Maiden Bower and Totternhoe to the border of Bucks. Documents dealing with property at Dunstable name it repeatedly from the twelfth century to this day.

The Romans were the first great makers of roads, here as elsewhere in Europe. Sections of Roman roads

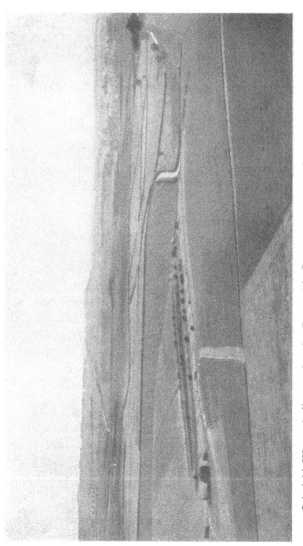

Icknield Way winding under the Dunstable Downs: source of the Ousel on the right

show sometimes three separate beds of varying kinds
built up beneath a surface of carefully-compacted stones
to a depth of more than three feet. When the earth
was by nature sufficiently solid they dispensed with
one or more of the lower beds. Mr Worthington Smith
records the exposure of 165 feet of the old Roman work
upon the Watling Street near Kensworth Lynch, between
the Horse and Jockey and Pack Horse inns. He describes the surface as made up of closely-compacted flint
stones, hard sandstone, and other stones such as are
found in the neighbourhood. A hole, picked with difficulty, showed a thickness of nine inches. Two Roman
roads entered the county. The Watling Street came in
between Caddington and Kensworth, and ran to the
Buckinghamshire border near Little Brickhill, crossing
the Icknield Way in Dunstable. The Stane (or Stone)
Way was a deviation from the Erming, or Ermine,
Street, the main north road from Colchester and London
to Lincoln and beyond. The Colchester Stane Way and
London Ermine Street met at Braughing, whence the
Stane Way passed through Baldock, Biggleswade, and
Sandy, then through or near Everton and Tetworth, and
so to Godmanchester in Huntingdon, where it again
joined the Ermine Street. A glance at the map will
show that this road follows almost exactly the course
of the Great Northern Railway, just as the London and
North-Western main line follows approximately that of
the Watling Street. The obvious explanation is that
like requirements produced similar results. The Roman
base of operations was a triangle, of which the three points

were Richborough and the neighbouring ports con-
centrating at Canterbury, London (the centre of trade
and key of the Thames valley), and Colchester (their
first military headquarters). From London and Colchester they pierced the country north-west and north,
to Chester on the one hand, and to Lincoln and beyond
on the other. The Chester road was carried along the
general line of the watershed, only crossing rivers
comparatively near their sources. The Lincoln road had
to deviate to the west to avoid the Fen country. The
Bedfordshire curve through Sandy may have been the
original route, or it may have been a later diversion
to avoid the greater extent of unsound soil at certain
seasons.

There are many evidences of the Roman origin of a
road other than those of its construction: the remains
of Roman camps, sites along the route, or place-names,
such as Streatford, Streatley, Stratton, Street, Chester,
Caster and its varieties, Cold Harbour and Caldecote,
Stone, Stane, Stanford, etc. It will often be found,
too, that a Roman road serves as a boundary; and
lastly there is the evidence of Roman military "road
books" or Itineraries. So the Watling Street is a
straight boundary between Tilsworth, Hockliffe, and
Heath-and-Reach (in the original parish of Leighton) on
the south, and Chalgrave, Battlesden, and Potsgrove on
the north; and also between Kensworth and Caddington;
while the Stane Way runs straight along the eastern side
of Stotfold, and continuing its line divides Astwick and
part of Biggleswade from Edworth. The rule is often

disregarded because parts of the Roman road had gone to ruin before parishes were formed, but such a "natural" division is effected by nothing else in Bedfordshire but Roman roads and rivers. The Stane Way has on it, too, such place-names as Stratton and Chesterfield, and the presence of Lower Caldecote on the other side of the Ivel makes it quite likely that the Way crossed at

Suspension Bridge, Bedford

Biggleswade, as the modern road does, and recrossed at Sandy. And the very names long persisted, for a deed of 1433 concerns property at Edworth "abutting on highway called Stanyway," and the Watling Street occurs as a boundary of property from the eleventh century onwards.

From the days of the Romans little serious road-making or even good repairing was done till the end of the eighteenth century. Under the Saxon kings the four great highways (quatuor chemini) were regarded as national roads. The "King's peace" extended upon them; violence and outrage were held to be more serious there than elsewhere. Their maintenance was a burden on the neighbouring landowners, and in 1285 the Prior and burgesses of Dunstable were ordered to repair the royal roads passing through their town. Later the Hundreds were probably responsible for the local roads. In the reign of Edward III the Commissions of the Peace were first issued, and the ancient machinery of County and Hundred Courts gradually fell into disuse. The parishes were held responsible for the highways within their borders, and they could be brought to do their duty by "presentment" at Quarter Sessions.

Of the details of movement upon these roads we only get chance glimpses. The third chapter of Macaulay's *History* gives a lively picture of English road-faring in the seventeenth and eighteenth centuries, and Bedfordshire was probably no exception. Pennant tells us that in 1740 the Chester stage did the journey from Dunstable to London in a day. The team consisted of "six good horses and sometimes eight." This meant starting two hours at least before daybreak during much of the year and often arriving late at night. He contrasts this with the improvements of forty years later. The direct road from Dunstable to Luton was made in 1784. Lysons speaks of the Bedford and Turvey road as out of repair, the Stagsden road as

Park Road, Luton

impassable for carriages, and most of the private roads as bad at the beginning of the nineteenth century.

Under a series of Acts passed in the eighteenth and nineteenth centuries many main roads had been given over to bodies of trustees, who were empowered to collect tolls from those who used the roads in order to provide a fund for their maintenance, as it was recognised that the localities through which the main roads passed did not supply any large fraction of the number of those who used them. The exaction of toll necessitated the use of the turnpike gates, and for the first time since Roman days, in England at least, scientific road-making attracted the attention of engineers. Under Telford and McAdam and their successors, gradient, curvature, material, and construction were considered and a new era of road-making began. But the toll system was found annoying and costly, and was abandoned, the last turnpike disappearing about twenty years ago. The Local Government Act of 1888 finally handed over the management of the main roads to the County Councils and that of 1894 transferred all other roads to District Councils, thereby abolishing the then existing Highway Boards.

In coaching days Dunstable saw more coaches than any part of Bedfordshire; the number is said to have risen to eighty a day. At the beginning of the nineteenth century the journey to London only took three hours and fifty minutes. In 1782 a new road was made to avoid the steep ascent of Chalk Hill between Dunstable and Hockliffe, and in 1837, when the threat of the coming railways had already reduced the number of coaches to twenty-eight a day, £10,000 was spent in cutting through Chalk Hill. But it was too late. The opening of the London and Birmingham Railway in 1838 brought Dunstable coaching almost to an end; and in November, 1846, on the opening of the Bletchley to

The "Bedford Times"
(One of the last coaches running from London)

Bedford branch of the London and North-Western Railway, the "Bedford Times" coach ceased to run.

The first of the great railway lines to connect London with England north of the Thames was the London and North-Western, and it provided Bedfordshire with a branch from Bletchley in 1846, and Dunstable with another from Leighton Buzzard in 1848. At that time the Midland had no line farther south than Leicester.

and made terms with the North-Western for the use of its line to London. In 1850 the Great Northern completed its line from King's Cross to Peterborough, and wishing to keep the Midland from a direct approach of its own to London, encouraged that company to continue its main line through Bedford to Hitchin, and gave it generous running powers and conveniences at King's Cross. This extension was opened by 1858. In 1851 the North-Western connected Oxford with Bletchley, and in 1862 a line was opened between Bedford and Cambridge, absorbing a small private line which Captain Peel had laid from Potton to Sandy. This the London and North-Western eventually took over. At last the Midland determined to continue their line direct to London, and in 1868 this was completed, the route through Luton being adopted. Luton was also joined to Dunstable by rail in 1858, and it is connected with Hatfield by a branch of the Great Northern. There is a branch of the Midland between Bedford and Northampton, and another from Kettering to Huntingdon skirts the extreme north of the county with a station at Kimbolton. But just as the two Roman trunk roads spread like open scissor-blades and left the north of Bedfordshire unheeded, so it was left by the Great Northern and North-Western, and even the advent of the Midland did little to fill the void.

The water communications of Bedfordshire have now lost their importance. The Grand Junction Canal just touches the county at Leighton Buzzard, but the railways which drove off the coaches destroyed the

Newnham Reach, River Ouse

barge industry as well. In the seventeenth century, when the French system of locks had been introduced into England, the Ouse was made navigable from King's Lynn to Bedford and the Ivel to Biggleswade, and for two centuries they continued to bring the bulk of the heavy merchandise that the north and centre of the county imported. But soon after the opening of the railway routes to the north and to London the river traffic decayed. About twenty years ago the locks were again put into repair and there was a modest revival of traffic, but it did not last long. Within the last few years proposals have been made to restore the trade by the use of steam or motor-driven barges, but they have not met with much support.

In connection with the rivers must be mentioned the repair and maintenance of the bridges. Mention of a bridge at Bedford occurs as early as 1188, when the Sheriff accounted to the Exchequer for money spent upon it. In 1258 we find record of a "mill by Biddenham Bridge," as Bromham Bridge was called till about a century ago. St Neots bridge is mentioned in the St Neots Cartulary as early as the first half of the thirteenth century, and Harrold Bridge already existed at the date of the Hundred Rolls of 1280.

Bedford and Biddenham bridges both had chapels attached to them. That at Bedford was endowed with several houses and shops, three acres of land, and a small rent from other lands. This endowment may have been partly for maintenance and repair of the bridge. The chapel was dedicated to St Thomas the

Martyr and was apparently "his chapel at Bedford Bridge" given by Simon de Beauchamp to St John's Hospital about 1180–90. The chapel on Biddenham bridge was built in 1296 and intended, according to its licence in mortmain, for the convenience of travellers,

Harrold Bridge and Causeway

to provide them with an early mass and to serve as a refuge from thieves. Some of our bridges are very long, spanning a wide extent of meadows liable to be flooded. Harrold bridge is for this reason approached by a long causeway pierced by culverts.

19. Administration and Divisions.

The Shire was a sub-division of the kingdom for three purposes—national defence, maintenance of law and order, and finance. From 1322 Bedfordshire had to provide small numbers of men for service against the Scots or French, but it was organised for its own pro-

Bedford County School or Elstow School
(*Founded in* 1869, *closed in* 1916)

tection as early as the Danish invasion at the beginning of the eleventh century. The general management of the county for the purposes of Justice, Administration, and Taxation, may be divided roughly into three periods ending with the Local Government Act of 1888, which

Bedford Grammar Scnool

(New Building opened in 1892)

established the present system of County Councils. Of
these periods the first ends with the Norman Conquest,
the second with the Act of 1360 which appointed
Commissioners (or Justices) of the Peace. Each county
was at first presided over by an *Ealdorman* (replaced
later by an Earl), but after the time of Canute a *Shire-
reeve* or Sheriff took his place, and the connection
between an Earl and a County became at last only
titular. Although "Hundreds" were older than Shires,
they formed a sub-division, and the Hundreds were
made up of townships or "vills." Every freeman was
bound to belong to a *tithing*, which, in Bedfordshire and
the neighbouring Midlands, was a group of ten men
headed by a tithing-man, not, as in some parts of
southern England, a sub-division of area. There were
Moots of the Hundred and Moots of the Shire. Meetings
or Moots of the Hundred were held at stated times,
and were attended by the parish priests and other
representatives of the vills. They assessed upon the
several townships their shares of the Hundred's proportion of the county-tax; made "*presentments*," or
reports, of nuisances, illegalities, and offences; and
settled such disputes as were within their competence.
Every year the Hundred was visited by the Sheriff, who
held his "view of frank-pledge," or inquiry as to whether
every man was in a tithing, so that someone might
be held answerable in case he offended. The moot or
meeting of the Shire was also held in the county town
at regular intervals, and in Saxon times it was presided
over by the Sheriff and Bishop of the Diocese. It

assessed upon the Hundreds their proportion of the
county taxes, sat as a court of jurisdiction, and had
presentments made to it by the Hundreds or by juries
who had been sworn to hold "inquisitions." It was
attended by the notables of the shire, and by the parish
priests and other representatives of the townships. The
Norman Conquest brought about one great change—
the bishop and clergy ceased to take part in the administration of the county. Otherwise the system described
went on for a couple of centuries. About 1200 a fresh
official—the Coroner—appears, with duties which remain
almost unaltered to this day. He still holds a "quest"
or enquiry, with a jury of neighbours, as to a death,
if its cause is unknown or suspicious, or if it has
occurred in prison, and as to any claims upon "treasure
trove." There were originally two coroners for Bedfordshire; now there is one for the county, a second for
the Honor of Ampthill (a collection of villages lying
mainly in Bedfordshire, but extending into Bucks), and
a third for the borough of Bedford.

Gradually the Hundreds did less business, and by
the time of Edward III the parish system had become
more generally organised. In 1360 the first Commission
of the Peace was appointed, and from then till 1888 the
Justices of the Peace were at once Magistrates and
Administrators of the Shire. Presentments of offences
were made to them at Quarter Sessions, as of old to the
County Court, and their Petty Sessional Divisions have
in some senses taken the place of the Hundreds. The
parishes supplied the machinery for the management of

highways and other purposes. The two most important functions of the Justices now are the administration of justice and, with the Standing Joint Committee, the management of the County Police.

Poor Law legislation began in the time of Elizabeth. Parishes have been combined in Unions, under the jurisdiction of Guardians. Such Unions are not necessarily confined within county limits; two-thirds, for instance, of the parishes in the Leighton Buzzard Union are in Bucks. The name of the County Courts is misleading. They have districts of their own, and are in no way connected with Administration, or with the shire. By recent legislation parishes may elect a Parish Council, which has certain powers of self-government, and some more populous parishes or areas have more considerable powers as Urban Districts. Kempston, for instance, is divided into Kempston Rural with a Parish Council, and Kempston Urban with a District Council. Leighton Buzzard, Ampthill, and Biggleswade are all Urban Districts.

Formerly Bedfordshire and Bedford were each represented in Parliament by two members, but by the Act of 1885 the borough lost one member. The county has two Parliamentary Divisions: the Southern or Luton Division is roughly all that part of the county south of Ampthill, the Northern or Biggleswade Division contains the rest of the county. Each of these divisions returns a member to the House of Commons, and the Borough of Bedford a third. No other town in the shire has ever been separately represented in Parliament except Dunstable, which returned two members in 1312.

The Hundreds have now little but historical importance. They are nine in number: Barford, Biggleswade, Clifton, and Flitt, whose names sufficiently suggest their position; Manshead on the west as far north as Aspley Guise; Redburnstoke, from Cranfield and Elstow in the north to Maulden in the south; Wixamtree, between Redburnstoke and Flitt on the west and Biggleswade on the east; Willey, mainly on the limestone in the north-west from Stagsden to Wymington; and Stodden, occupying that part of the county north of the Ouse unoccupied by Willey or Barford.

The County Council took over the administration of the County in April 1889. There are sixty-eight members, elected by occupiers of buildings or land in the county, irrespective of sex. One-fourth of the members are Aldermen, who are elected by the other Councillors. The officials of the County Council include a Clerk (who is also Clerk of the Peace for the County), a Treasurer, Analyst, Surveyor, Secretary for Education, and Inspector of Weights and Measures. The County Coroner is elected, but the Coroner for the Honor of Ampthill is appointed by the Crown. The general business of county administration is in the hands of the Council, and concerns highways and bridges, education, lunatic asylums, contagious diseases (animals), pollution of rivers, parliamentary registration, and finance.

The Justices hold their Court in general sessions four times a year at least (Quarter Sessions), and oftener if they think fit. They try criminal cases involving very serious penalties; but murder, bigamy, treason, forgery, and some others of the graver offences must go before the Judges of Assize. They also sit in Petty Sessions as courts of summary jurisdiction, to try minor offences without a jury. For this purpose the county is divided into seven divisions, severally named after the chief

Town Hall, Bedford; formerly the Grammar School

place in each: Ampthill, Bedford, Biggleswade, Leighton Buzzard, Luton, Sharnbrook, and Woburn. The Justices of the Peace also act as Licensing Authorities within their petty sessional divisions.

The Poor Law Guardians have the control, under the direction of the Local Government Board, of all matters concerned with the administration of the Poor

Law within the Union of Parishes which they represent. To the tax upon each parish levied for the purpose have been gradually added the taxes required for a number of other local purposes, with the result that the so-called Poor Rate is really a rate to meet the requirements both of the Poor Law authorities and other provinces of county administration with which the Poor Law guardians have nothing to do. A town with 50,000 inhabitants may be constituted a County Borough, and is then independent of the administration and finance of the County Council, but there are none such at present in Bedfordshire.

The Borough of Bedford has its own Coroner and its Justices of the Peace. When the latter sit in Quarter Sessions the Recorder of the Borough, appointed by the Crown, is the Judge of the Court. For general purposes of administration Bedford, Luton, and Dunstable are Municipal Corporations, consisting of Mayor, Town Councillors elected by the ratepayers, and Aldermen elected by the Councillors.

20. The Roll of Honour.

Bedfordshire, though not a large county, can claim a long list of distinguished names, of which a few only can be given here. We may begin with the clergy.

Peter Bulkley, who succeeded his father in 1620 as Rector of Odell, sailed for Boston in Massachusetts in 1635 and founded the township of Concord. He was its first minister, and his son one of the earliest Harvard graduates. John Donne the poet (1573–1631) was Rector of Blunham for some time and the living was not a sinecure, though he held it with the Deanery of St Paul's. Isaac Walton has described his excellence as a preacher and Ben Jonson held him to be "the first poet in the world in some things." A younger contemporary, but of a very different type, was William Dell, a Bedfordshire man by birth, and Rector of Yelden. He served as chaplain in Fairfax's army (1645–7), and was appointed Master of Gonville and Caius College, Cambridge, but will be most remembered as having invited John Bunyan to preach in his pulpit at Yelden. At the Restoration he lost his Mastership and his Rectory, and when he died four years later (1664), he was, by his own wish, buried in unconsecrated ground.

Edward Stillingfleet (1635–99), who owed his nickname of "The Beauty of Holiness" to the comeliness of his face and character, was not a Bedfordshire man, but became Rector of Sutton in 1657. Here he wrote his famous *Origines Sacrae* and the *Eirenicon*, a "Message of Peace," which, by the irony of circumstances, was published (1659) just before the actual arrest of John Bunyan at Lower Samsell.

The lawyers and judges connected with Bedfordshire were generally either local magnates who were employed in early days by the Crown as "Justices Itinerant"; or successful lawyers who became possessed of estates in the county; or natives of Bedfordshire who rose to legal eminence. The Beauchamps, Cantilupes, Braybrooks, and Patteshalls afford examples of the first type in the twelfth and thirteenth centuries.   William de Beauchamp of Bedford was a Baron of Exchequer under Henry III.   It was the seizure of Henry de Braybrook after the Dunstable Sessions that led to the ruin of Fulk de Bréauté and the destruction of Bedford Castle in 1224; and one of the Justices who had been sitting with him, but avoided capture, was Walter de Patteshall of Bletsoe, an ancestor both of our present King and of Lord St John of Bletsoe.

Of the second type was William Inge, who in the thirteenth century acquired land in many counties. Among his manors was that of Weston, which has been called Weston Ing ever since.   John Cockayne, too, Baron of Exchequer and Justice of Common Pleas in the fifteenth century, bought Hatley, known till then as Hatley Port, but since his time as Cockayne Hatley, and in the same century a son of the celebrated Chief Justice Gascoigne acquired Cardington by marrying a Pigott heiress.

Sir Robert Catlin, Chief Justice of the Queen's Bench (1562–74) was born at Sutton, and died at the house he had built upon the site of Newenham Priory. The ruins of this house still stand upon the sewage-farm field.   Among his descendants were Marlborough and the Spencers of Althorp.   Sir Francis Crawley, a Justice of Common Pleas, was the head of a family long since, and still, settled at Luton.   In 1636 he supported the right of the Crown to levy ship-money, and he was one of the Judges who decided against Hampden.   He married a Rotheram, and so probably acquired Someries, near Luton, where he lived till his death in 1649.

Oliver St John of Keysoe (1578–1673), a cousin of the St Johns of Bletsoe, was doubly connected by marriage with Cromwell and Hampden. He was Hampden's counsel in the ship-money case, and a strong supporter of the bill of attainder against Strafford. He became Chief Justice of Common Pleas in 1648. Though believed to be in Cromwell's confidence, he opposed the proceedings against the King. He died in 1673. Sir Samuel Brown, of Arlesey, was Oliver St John's first cousin. He was a Justice of the King's Bench, but resigned that office when the King was brought to trial.

Sir George Carteret held Jersey for the King in the Civil War, became Treasurer of the Navy under Charles II, and bought the Haynes estate from Sir Samuel Luke in 1667. He was buried there in 1679 and two years later his grandson George was made Baron Carteret of Haynes. Lord Carteret was a remarkable man. After a youth of eager study at Westminster and at Oxford, he entered early into political life, served as Ambassador in Sweden and as Secretary of State for Foreign Affairs, and was Lord Lieutenant of Ireland. But the detail of his offices gives no measure of his powers, and very many of his contemporaries testify to his abilities. "When he dies," said Chesterfield, "the ablest leader in England dies."

Sir Samuel Luke, a Puritan landowner of Cople and Haynes, had Samuel Butler in his service, probably in the capacity of clerk or secretary, and is said to be the original of *Hudibras*. But, as many caricatures do, that amusing poem, which was written at Cople, appears to have exaggerated his physical disadvantages and foibles, and ignored his real qualities. He was for some time in command of the important post of Newport Pagnell, and showed great energy and courage. When the war was over he lived at Haynes, and Dorothy Osborn writes from Chicksands in 1653 "Of late, I know not how, Sir Sam hath grown so kind as to send to me for some things he desired out of this garden, and withal made the offer of what was in his, which I had reason to take for a high favour, for he is a nice florist."

John Okey the Regicide was a London drayman who obtained the lease of Ampthill and Brogborough Parks, and represented the shire in Richard Cromwell's parliament. He fled to Holland at the Restoration, was arrested, sent to England, and hanged at Tyburn in 1662.

John Howard, the philanthropist (1726–1790), though not actually Bedfordshire born, spent his childhood at his father's house at Cardington. When on a voyage his vessel was captured by a French privateer, and Howard experienced the miseries of a dungeon at Brest, an experience destined to bear fruit later. Returning to England he made Cardington his home, and became High Sheriff of Bedfordshire in 1773. His official visits to the jails drew his attention to their abuses, and he devoted the remaining years of his life to a careful accumulation of facts illustrative of the conditions of prisoners, not only in Great Britain but throughout

Statue of John Howard, Bedford

Europe. His private life at Cardington was devoted to philanthropy. He rebuilt every cottage upon his estate, had his household linen woven by the cottagers, and introduced new kinds of potatoes. "I am a plodder," he said, "who goes about to collect materials for men of genius to make use of." But Burke has judged him otherwise:—"as full of genius as of humanity."

Samuel Whitbread (1720–1796) owned considerable estates at Cardington, was member for Bedford for many years, and by his will he founded the Bedford Infirmary. His son, Samuel Whitbread junior (1764–1815), was also member for Bedford, from the year 1790 till his death. His grandson—also of the same name—successfully contested Bedford in no less than eleven elections, and held his seat till his resignation. He was connected with all the modern developments of the borough.

James Howard (1821–1889) was of a family settled in Bedford for at least two centuries, the son of an agricultural implement maker in the High Street. He built the "Britannia" works on the site of Caldwell Priory in 1856. He represented Bedford in Parliament from 1868 to 1874 and the county from 1880 to 1885, helped to found the Farmers' Alliance, and was its President for many years. He used his Clapham estate for experiment, and published many of his results, and as long ago as 1869 tried to attract public attention to the manufacture of beetroot sugar in England.

Sir Joseph Paxton (1801–1865) was born at Milton Bryant, the son of a small farmer, and educated at the Woburn grammar school. He began his career as a boy

Sir Joseph Paxton

in the gardens of Sir Gregory Page-Turner at Battlesden and gave evidence of his abilities by laying out and constructing the large lake there. He attracted the attention of the Duke of Devonshire and was appointed gardener at Chatsworth, where he built the great conservatory. In 1851 he designed the vast glass building

Sir William Harpur

of the Great Exhibition in Hyde Park, which was afterwards re-erected at Sydenham and is known throughout the world as the Crystal Palace. He was knighted and became M.P. for Coventry.

Sir William Harpur (1506–1573) was born at Bedford, served as Sheriff in 1556–7, as Lord Mayor of London in

1561–2, and was knighted during his mayoralty. He possessed, probably by inheritance, the land at the west end of St Paul's Square, on which the Town Hall now stands, and there, probably in 1548, he built a school house. In 1552 Edward VI granted a patent for the esta-blishment of a free and perpetual school in Bedford. In 1564 Sir William bought 13¼ acres of land in Holborn for £180, and two years later a deed between Sir William Harpur and his wife Alice of the one part, and the Corporation of Bedford of the other, attests the foundation by the Corporation of a free school in Bedford, in the school house "lately built by Sir W. Harpur," the appointment of Edmund Green of New College as its first master, and its endowment with the land and building then in use and the 13¼ acres of land in Holborn. The latter then let at about £12 : the income now received from the Bedford and London properties of the Harpur Trust amounts to £18,000.

Among famous women must be mentioned the daughter of Margaret Beauchamp, the heiress of Bletsoe, Lady Margaret Beaufort, who married Edmund Tudor, and whose son eventually became King as Henry VII. She was a woman of remarkable character and ability. She translated the *Imitatio* from the French, patronised Caxton and the early English printers, founded Professorships in Divinity at Oxford and Cambridge, a school at Wimborne, and the Colleges of Christ and St John at Cambridge.

At Houghton House, whose ruins are still so familiar, lived Mary, the sister of Sir Philip Sidney and the mother of William Earl of Pembroke. Francis Osborn, a Bedfordshire man, who served for a time as Master of the Horse to her son, speaks of her wit and beauty, and quotes Ben Jonson's famous epitaph:—

> Sidney's sister, Pembroke's mother,
> Death! ere thou hast slain another
> Learn'd, and fair, and good, as she,
> Time shall throw a dart at thee.

Chicksands Priory

While Francis Osborn was busy at Oxford writing political tracts that have been long forgotten, his niece, Dorothy Osborn, was tending his brother's last years at Chicksands, and writing to her lover the wonderful letters that have made every reader her friend. Seventy-seven letters which she wrote to Sir William Temple between 1652 and 1654 contain a weekly record of her life. They depict with charming and artless verve her devoted attendance upon her father, her resolute loyalty to her lover in face of her brother's opposition, whimsical accounts of other suitors, her intercourse with neighbours and friends, her delight in those long French romances which were so fashionable in her day, and the quiet charm of her country home, with its garden and river, and the common beyond it. The house still stands, a little altered, the common is now enclosed within the park, but keeps its name, and in the garden, between it and the common, still flows the little river, and its blue forget-me-nots appeal to us to remember Dorothy Osborn.

Catherine of Aragon lived at Ampthill Castle for about a year, leaving before or in the early part of 1534. The divorce was pronounced at Dunstable in May 1533, and in May 1534 she was already at Kimbolton, where she died in 1536.

Among statesmen Bedfordshire can claim more than one illustrious example in the last four centuries; the Russell family being especially conspicuous. "Imagine always," writes Sir Thomas Wyatt, "that you are in the presence of some honest man that you know, such as Sir John Russell, and remember what shame it were afore him to do naughtily"; and he goes on to define the honesty of which he speaks as "wisdom, gentleness, soberness, desire to do good, friendliness to get the love of many." It was not till late in his career that John Russell, Earl of Bedford (*c.* 1486, *d.* 1555), who rose rapidly to high offices of state under Henry VIII, and was a man of very fine character, became closely associated with

William, first Duke of Bedford

Bedfordshire, though he had long since married the widow of Sir John Broughton of Toddington. Henry VIII made him a Baron: it was in the reign of Edward VI that he obtained the grant of Woburn Abbey and the Earldom of Bedford.

His great grandson, Francis, the fourth Earl of Bedford (1593–1641), held so important a position in the eyes of both King and Parliament that Clarendon regards his death as one of the immediate causes of the Civil War. One great service he did to the country raised an imperishable monument to himself. He was chief "undertaker" of the scheme for draining the Fens, and the work was brought to completion by his son William, the first Duke, late in the century. The "Bedford Level" commemorates the part they took in the achievement.

William, Lord Russell (1639–1683), third son of the fifth earl, was obnoxious to the Court because of his uncompromising support of the Exclusion Bill. As an accomplice in the Rye House Plot he was arrested, and though no complicity was proved against him he was executed 21st July, 1683. The sympathy and apprehension aroused by his death contributed not a little to the fall of the Stuart dynasty.

John, fourth Duke of Bedford (1710–1771), took a conspicuous part in public affairs during the middle years of the eighteenth century, as First Lord of the Admiralty, Lord Lieutenant of Ireland, Secretary of State, and Ambassador to France. He was an industrious administrator, an able and an honourable man, "a man of inflexible honesty and love for his country," though his reputation still suffers from the perverse imputations of the *Letters of Junius*.

John Russell, first Earl Russell (1792–1876), was doubly a Bedfordshire man. His father, Lord John Russell, was brother of the fifth Duke of Bedford, and his mother was a Byng, daughter of Lord Torrington of Southill. Though born in London, Woburn Abbey was his home for nearly twenty years. He entered Parliament in 1813 and became an ardent champion of Parliamentary reform. From 1820 to 1832 he lived at the Hundreds Farm at Woburn, where he had his fine library. He had charge of the Reform Bill in the House of Commons (1830-2), was Prime Minister 1846–52, became Earl Russell 1861, retired from political life in 1866, and died 1876.

Two Bedfordshire men at least have earned fame in war. Sir Nigel Loring, who came of an old family settled at Chalgrave, was a comrade of the Black Prince, and one of the original members of the Order of the Garter He fought at the siege of Calais, and at the battle of Poictiers. There are in Chalgrave Church two figures of Lorings as knights in armour, and one of these is undoubtedly his. Admiral George Byng of Southill won the naval victory over the Spaniards off Cape Passaro in 1718, and was created Baron Southill and Viscount Torrington. It was his son, John, who was the famous and unfortunate Admiral Byng who was shot in 1757 for failure to save Minorca. Both are buried at Southill, and the inscription over the son's tomb begins:—" To the Eternal Disgrace of the British Nation..."

The county has been neither the birthplace nor the home of any of the greater English poets, though it can claim some of minor note. George Gascoigne (*d.* 1577), of Cardington, was M.P. for Bedford from 1557 to 1559, and published numerous lyrics, elegies, and other verses and translations. Sir Sidney Lee points out that we owe to him the earliest extant comedy in English prose, and the earliest English critical essay on versification. Bedfordshire has two poets laureate of no great distinction. Elkanah Settle (1648–1724), born at Dunstable, where his father was a barber and innkeeper, held the post of Poet to successive Lord Mayors for many years, devised their pageants and shows, and died in the Charterhouse. Nicholas Rowe (1674–1718) was born at Little Barford. The *Fair Penitent* and *Jane Shore* are the best known of his plays, and we owe to him the earliest critical edition of Shakespeare.

John Bunyan's fame is such that it is not necessary to do more than note the epochs of his career. He was born at the east end of the parish of Elstow where it abuts upon Harrowden, in the fields of the farm still known as Bunyan's End. The cottage stood on the north side of the stream, but on ground which is now on the south of it, as the brook has been straightened: the field north of the stream is still called Pesselynton, as it was even before his day. His family were already living there in 1327, and the name occurs in the neighbouring village of Wilstead in 1199. He was baptised at Elstow, 30th November, 1628. His father was a brasier, and he was brought up to the same trade. At the age of sixteen he joined Col. Cockayne's Company of the Newport Pagnell garrison under Sir Samuel

Statue of Bunyan and St Peter's Church, Bedford

Luke. In the muster rolls, certified by Henry
Whitbread, muster-master, his name occurs for three
years and a half, beginning with 30th November, 1644.

John Bunyan

He married about the time at which his military period
ended (1648) and settled as a brasier at a cottage in the
High Street of Elstow of which the present "Bunyan's
Cottage" is a reconstruction. The spiritual struggle

narrated in *Grace Abounding* belongs to his life in
that cottage. In 1650 the congregation now known as
the Bunyan Meeting was formed under the ministry of
John Gifford. In 1653 the Corporation presented John
Gifford to the living of St John's Church, episcopal
ordination not being at the time necessary. In 1655
Bunyan joined this congregation, and he removed to
Bedford in 1654 or 1655. His first book was published
at Newport Pagnell (1656) and arose from a controversy
with Quakers. He was soon known as a preacher in
Bedford and the neighbourhood, and was invited by
William Dell to preach in his pulpit at Yelden on the
Christmas Day of 1659, the year in which he married
his second wife. On 12th November, 1660, he was
arrested for unlicensed preaching at Lower Samsell in
Westoning parish, and the place of his arrest may still
be identified in a field between Lower Samsell and Harlington. He was committed to the County Gaol which
then stood on the north of the open space at the High
Street end of Silver Street, Bedford. His imprisonment
was sometimes lax and sometimes strict, and lasted till
the publication of the Act of Indulgence, March 1672.
During that period he published *Grace Abounding* and
eight other books. In January, 1672, Bunyan was
elected minister of the Congregation, and in the following
May a barn in an orchard in Mill Street was licensed as
a place of Congregational Meeting. On the site of that
barn the present Meeting House stands. Bunyan was
again imprisoned for about a year, from 1675 to 1676, by
the borough authorities of Bedford in the town gaol

upon the bridge, and it was there that he wrote the first part of the *Pilgrim's Progress*, which was published in 1678. He preached his last sermon in London, 19th August, 1688; died, after ten days' illness, on the last day of the same month, and was buried in the cemetery at Bunhill Fields.

21. THE CHIEF TOWNS AND VILLAGES OF BEDFORDSHIRE.

(The figures in brackets after each name give the population in 1911, and those at the end of each section are references to the pages in the text.)

Ampthill (2270), 8 m. S. of Bedford, on the sand, with stations on M.R. and L. and N.W.R., is an Urban District. The Castle, now destroyed, was the residence of Catherine of Aragon during the divorce proceedings. Ampthill House, built in 1694, has a lime avenue, the Alameda, reputed to be one of the finest in England, and famous oaks of great age, some measuring over 36 feet in circumference. Church (St Andrew) mainly Perp. (pp. 18, 24, 32, 34, 60, 65, 67, 73, 83, 118, 127, 145, 146, 147, 148, 153, 160.)

Arlesey (2046), 4 m. S. of Biggleswade, a straggling village some two miles in length with G.N.R. stations at each end, has engineering, Portland cement, and large brick and tile works. The Eltonbury earthworks are in the neighbourhood. Church of St Peter remarkable for its massive western tower, built in 1187. (pp. 70, 72, 152.)

Aspley Guise (1277), 2 m. N. by W. of Woburn, stands high (about 400 ft.) on the sand, among the firs, and is a health resort. The church (St Botolph), partly Dec., has been recently restored. (pp. 12, 127, 147.)

Barford, Great (726), a large village on the Ouse 5½ m. N.E. of Bedford, has a fine 15th cent. bridge of 17 arches.

Barford, Little (151) on the Hunts border 3½ m. S. of St Neots. Church (St Denys) has a Norman arch to S. door. In a cottage still standing Nicholas Rowe was born in 1674. (pp. 103, 147, 164.)

Barton-in-the-Clay (746), 6 m. N. of Luton, stands just north of the chalk escarpment half-way between Luton and Ampthill. It once belonged to Ramsey Abbey. The church (St Nicholas) has a fine Perp. roof of chestnut wood, but is chiefly E.E. and Dec. It contains some early brasses, and the chancel is paved with 14th cent. tiles. (pp. 12, 15, 16, 63, 71.)

Battlesden (66), 3½ m. S. of Woburn. Church (St Peter) is chiefly Perp. Sir Joseph Paxton was born here. (p. 132.)

Bedford (39,183) the county town, 49 m. from London on the Ouse, is served by the Midland and London and North Western Railways. Is a municipal and parliamentary borough, head of a petty sessional division and county court district. The town lies mainly north of the river, and though showing few signs of antiquity is bright, clean, and well planted. The Ouse was once navigable to King's Lynn, but is so no longer, though there is much boating, and regattas are held. There are some interesting churches. St Peter's has Saxon details in the tower, a Norman S. door, and still shows traces of having been set on fire by the Danes in 1010. St Paul's was rebuilt in the 13th cent., and has been much altered at many subsequent periods. St Mary's is mainly Perp. Behind the Swan Inn is the site of Bedford Castle, destroyed in 1224. Bedford Grammar School dates from 1552, and is now one of the leading Public Schools, owing its prosperity in great measure to the benefactions of Sir William Harpur its founder. In 1853 the Commercial School was separated from it and was re-named the Modern School in 1877. The High and Modern Schools for Girls were established on the same foundation in 1882.

There are large iron-works (built partly on the site of Caldwell Priory), agricultural implement works, automobile and various other manufactories, and breweries, and pillow lace is still made. From Bedford eastward to the site of Newenham or Newnham Priory the river banks have been laid out with gardens and ornamental walks. On St Peter's Green is the bronze statue of

Bunyan by Boehm, presented by the Duke of Bedford, and Gilbert's statue of John Howard stands on Market Hill. (pp. 2, 3, 4, 6, 8, 11, 17, 21, 22, 27, 28, 44, 45, 53, 58, 65, 67, 70, 72, 74, 75, 76, 77, 78, 79, 80, 81, 82, 83, 84, 85, 86, 91, 92, 97, 98, 99, 101, 103, 111, 113, 115, 118, 119, 126, 127, 128, 133, 134, 137, 138, 140, 141, 142, 143, 145, 146, 148, 149, 151, 154, 155, 157, 158, 164, 165, 167.)

The Modern School, Bedford
(The old Grammar School and the Girls' Modern School in the distance)

Biddenham (451), on the Ouse 2 m. W. of Bedford, is chiefly noteworthy for the finds of extinct mammalia in its gravels, which have also yielded a number of palaeolithic implements. The church of St James has a Norman chancel arch. (pp. 22, 68, 70, 91, 94, 96, 140, 141.)

Biggleswade (5375), a town 10½ m. S.E. of Bedford, on the Ivel, in flat country much devoted to market gardening. There are fairs at Easter and on Whit Monday. (pp. 6, 12, 18, 53, 54, 67, 70, 131, 132, 133, 140, 147, 148.)

Bletsoe (312), a village 6 m. N.N.W. of Bedford, by the Ouse. The cruciform church of St Mary is chiefly of Dec. period, and contains some fine monuments of the St John family. A portion of Bletsoe Castle still remains as a farm house. Pillow lace is made. The Falcon Inn was a favourite resort of Edward FitzGerald. (pp. 82, 83, 126, 127, 151, 152, 158.)

Blunham (603), a large village on gravel soil on the Ivel near its junction with the Ouse, 7 m. E. of Bedford. The church (St James), which belonged to St Edmund's Abbey, is partly Norman. John Donne was once its Rector. There is a good deal of market gardening. (p. 150.)

Bolnhurst (184), a small village 7 m. N. by E. of Bedford, is chiefly noticeable for its church (St Dunstan) which has a fine carved screen and a mural painting of St Christopher. (pp. 11, 16, 123.)

Bromham (350), an attractive village 2½ m. N.W. of Bedford, on the Ouse. The Hall was held by the Royalists and captured by the Parliamentary troops. The church (St Owen) built in 13th cent. has some interesting monuments. (pp. 17, 22, 34, 91, 92, 96, 140.)

Caddington (1508), a village on the outskirts of Luton till lately partly in Herts, with an interesting church (All Saints') showing Norman, E.E., and Perp. work. In the neighbourhood is an ancient British camp. (pp. 9, 14, 15, 27, 34, 72, 93, 96, 131, 132.)

Campton (415), a small village 6 m. S.W. of Biggleswade where many Anglo-Saxon and Roman antiquities have been found. The Manor House was attacked by the Parliamentarians in 1645. Robert Bloomfield the poet is buried here. (pp. 113, 121, 122.)

Cardington (423), a well-wooded and pretty village 2½ m. S.E. of Bedford, once the residence of John Howard the philanthropist. The church, mainly modern, has a Norman arch in

the tower.   Cardington Cross was the work of Chantrey.
Straw-plait and pillow-lace making, once large industries, have
almost ceased to exist. (pp. 94, 120, 122, 123, 151, 153, 155, 164.)

Carlton (314), **Chellington** (113). Contiguous villages
9 m. N.W. of Bedford on the Ouse. The churches of both
show interesting E.E. and Dec. work. (pp. 111, 113, 115.)

Chalgrave (573), 3½ m. N. of Dunstable, has an ancient
church (All Saints') of various periods, with two good altar
tombs. (pp. 99, 108, 132, 163.)

Chellington (113), *see* **Carlton**.

Chicksands Priory (39) is now a parish, annexed to
Campton. The Gilbertine priory was founded about 1150, and
much remains of the original buildings, notably the cloisters
of the quadrangle. (pp. 125, 153, 159.)

Clapham (748), a village 2 m. N.W. of Bedford, remarkable for its church of St Thomas à Becket, which has a celebrated
Saxon tower 82 ft. in height. The two lower sections are very
early work, the third somewhat later. The main part of the
church was rebuilt in 1861 by Sir Gilbert Scott. (pp. 70, 92,
98, 101, 115, 117, 155.)

Cockayne Hatley (110), a little village on the extreme
E. border of the county 2½ m. E. of Potton, has a 14th cent.
church, which is especially noteworthy for its profusion of
beautiful carved woodwork of the stalls and chancel, etc.,
brought from the Abbey of Alne, near Charleroi. (pp. 117, 151.)

Cople (377), a village 4 m. E. of Bedford, with a good
church (All Saints') mainly E.E. and Perp. with two altar tombs
and several brasses. Samuel Butler wrote his *Hudibras* at Wood
End House. (pp. 96, 115, 152, 153.)

Cranfield (1199), a large village on the Bucks border 8 m. S.W. of Bedford, with an E.E. and Perp. church, dedicated to

Dunstable Priory

St Peter and St Paul, which once belonged to Ramsey Abbey. (pp. 16, 44, 96, 103, 147.)

Dean (342), a village at the extreme N. of the county, 4 m. W. of Kimbolton, divided into Nether and Upper Dean. The church (All Hallows'), chiefly Dec. and Perp., has a very fine carved roof.

Dunstable (8057), a large market town and municipal borough in the south of the county 5 m. W. of Luton, on L. and N.W.R. and G.N.R. It stands on the site of a Roman town where Watling Street crosses the Icknield Way.   Henry I founded here a Priory of Augustinian Canons, a portion of which still remains as the parish church of St Peter, which shows very fine Norman work in the nave and W. doorway, in addition to the rich E.E.   Many kings and great officials stayed in the town.   Before the days of railways it was a great coaching centre, and at one time was the headquarters of the straw-pla industry.   There are now large paper and engineering works, lime works, etc.   Dunstable is on the edge of the chalk and in the neighbourhood on the downs are many relics of prehistoric man; notably "Maiden Bower," a British Camp, 10 acres in area, surrounded by a circular vallum about 10 ft. high; and five round barrows, as well as hut circles.   Elkanah Settle was born here in 1648.   (pp. 6, 7, 9, 12, 14, 15, 18, 34, 59, 60, 61, 64, 70, 71, 74, 78, 79, 80, 81, 82, 84, 95, 96, 103, 104, 105, 108, 126, 129, 130, 131, 134, 136, 137, 138, 149, 151, 160, 164.)

Eaton Bray (979), a village 4 m. W. of Dunstable, has a good church (St Mary the Virgin) with E.E. nave arcades and font, and very fine 13th cent. ironwork in the S. door, recalling that of Queen Eleanor's screen in Westminster Abbey.   (pp. 9, 55, 105, 107, 108, 109.)

Eaton Socon (2319) close to St Neots, on the Hunts border, lies on the Ouse, which is here crossed by a bridge built in 1589 from the ruins of St Neots Priory.   The parish comprises a number of hamlets, of which Bushmead still shows the refectory of its ancient priory (Augustinian). The church of Eaton Socon (St Mary) is mainly Perp. with a good rood-screen. There is much market gardening. (pp. 4, 9, 11, 16, 52, 99, 113.)

Elstow (499), a village 1 m. S of Bedford. John Bunyan was born here in 1628. The church is part of the nave of a

Bunyan's Cottage, Elstow

Benedictine nunnery and is very interesting, containing early Norman work, some fine E.E., and a noteworthy brass of the last Abbess, Elizabeth Hervey. On the village green is a large half-timbered building and ruins of a Jacobean house adjoin the church. (pp. 96, 103, 110, 115, 124, 126, 147, 164, 166.)

Felmersham (345), a village on the Ouse 7 m. N.W. of Bedford, notable for its church (St Mary), perhaps the most interesting in the county, a cruciform E.E. building with fine central tower and a remarkable Perp. rood-screen.   (pp. 21, 94, 106, 108, 109, 113, 115, 117, 128.)

Flitton (463), a village 2½ m. S.E. of Ampthill, on the Flitt, is, with its hamlet Greenfield, mainly devoted to market gardening. Its church of St John Baptist, late Perp., is noteworthy for the attached mausoleum of the De Grey family with fine monuments dating from the 16th cent. (pp. 96, 113.)

Flitwick (1424), 2½ m. S. of Ampthill, on the Flitt.   The church has a Norman doorway.   Ruscox manor has been bought by the County Council for small holdings. (pp. 6, 18, 19, 32, 53, 68, 99.)

Goldington (967), a village on the eastern suburbs of Bedford, once the site of Newenham Priory. Near by is " Risinghoe Castle," a high mound, probably the remains of a *burh*, dominating the river.

Gravenhurst (377), a civil parish comprising Upper and Lower Gravenhurst, lying some 12 m. S.E. of Bedford.   At the former St Mary's church, mainly Perp., has a good Norman chancel arch, and in a church with similar dedication at the latter there is a Dec. rood-screen and an early brass to Sir Robert de Bilhemore. (pp. 34, 72.)

Harrold (851), a village 10 m. N.W. of Bedford, on the Ouse, with an interesting church, Trans. Norman, E.E. and Dec., and an octagonal market-house. (pp. 11, 22, 67, 108, 140, 141.)

Haynes (Hawnes) (676), a village 6 m. S.E. of Bedford, chiefly noteworthy from the seat, Haynes Park, which in 1667 came into the possession of Sir George Carteret the Royalist. (pp. 96, 122, 152, 153.)

Higham Gobion (76), a small village on the Herts border, once the residence of Stephen Castell, the learned author of the *Lexicon Heptaglotton*. Many Roman antiquities have been found in the neighbourhood. (pp. 82, 96, 111.)

Houghton Conquest (535), a village 3 m. N. of Ampthill; the church has a very large fresco of St Christopher. (pp. 83, 123, 158.)

Houghton Regis (1369), a large village 1 m. N. of Dunstable, with a good Dec. church containing a fine Norman font. The straw-plait industry is still carried on. (p. 15.)

Husborne Crawley (379), 1½ m. N.E. of Woburn. The experimental farm of the Royal Agricultural Society is mainly in this parish. (p. 51.)

Hyde (649), a civil parish formed in 1896, partly from Luton, with additions in 1906 of portions of Harpenden in Herts. The fine seat, Luton Hoo, is in the parish, and Someries (1448), of which the gateway and some other remains still exist.

Kempston (5349), a large parish and village contiguous with Bedford on the S.W. with an increasing population. Here, in the extensive gravel pits, many palaeoliths have been found and, some years ago, large Romano-British and Anglo-Saxon cemeteries which yielded a great number of relics. The church (All Saints') is Norman with a Dec. nave. (pp. 18, 44, 55, 68, 92, 94, 97, 98, 103.)

Kensworth (528), a small village about 2 miles S. by E. of Dunstable, of which the greater part was in Herts until 1897. The church of St Mary shows a good deal of Norman work. (pp. 9, 12, 14, 15, 26, 27, 42, 93, 96, 103, 131, 132.)

Knotting (120), a village on the Northants border 10 m. N. by W. of Bedford, chiefly of interest for its church of St Margaret, parts of which are Norman, the remainder of nave and chancel E.E. (pp. 11, 103.)

Leagrave (1270), 2½ m. N.W. of Luton, of which it is practically a suburb. Waulud's Bank, an old British camp, is here, and the Icknield Way crosses it. (pp. 14, 74, 95, 129.)

Market Cross, Leighton Buzzard

Leighton Buzzard (6782), an old town on the Bucks border, 20 m. S.S.W. of Bedford, on the Ousel, with large stock and wool fairs. The chief industry is sand and gravel digging. The church (All Saints') is cruciform and mainly E.E., with a fine tower and spire. The market cross, dating from the beginning of the 14th cent., is now a good deal restored. Many Saxon antiquities have been found at Dead-man's Slade in the neighbourhood. There are some modern engineering works. (pp. 3, 9, 11, 18, 68, 84, 109, 111, 115, 116, 137, 138, 146, 148.)

Luton Parish Church

Limbury and **Biscot** (2242), once hamlets of Luton, were formed into a parish in 1896. Limbury may be the "Lygean-burh" of the Anglo-Saxon Chronicle. "Bishopscote" was conferred by King Offa on St Alban's Abbey in 791. Various Roman and pre-Roman remains have been discovered. (p. 74.)

Luton (49,978), a large municipal borough in the south of the county near the Herts border, 31 m. from London, and 9 m

Canopied Font, St Mary's, Luton

S.W. of Hitchin. It lies on the Lea in a hollow surrounded by low chalk downs, a busy town, the largest in the county, the centre of the straw-hat industry, and extensive dye works in connection with it, as well as various factories. The cruciform church of St Mary, one of the finest in the county, has much good E.E. work, and a remarkable canopied 14th cent. font of great interest. There are also some good tombs. The Plait Hall, for the use of the plait dealers, is recent. (pp. 4, 7, 9, 14, 44, 53, 59, 60, 62, 63, 64, 65, 66, 67, 71, 72, 74, 75, 80, 81, 96, 97, 117, 126, 127, 128, 134, 135, 138, 146, 148, 149, 151, 155.)

Marston Moreteyne (1025), a village 4 m. N.W. of Ampthill, of interest for its church (St Mary), E.E. and Perp., with its very massive detached tower. There is a moated manor house near. (pp. 115, 121, 123.)

Meppershall (610), a village about 4 m. N.W. of Hitchin. The church (St Mary) stands on high ground and has a very extensive view. It is cruciform, and the tower and part of the transepts are Norman. St Thomas's Chapel about 1 mile distant, now used as a barn, has a Norman doorway and two elegant Dec. windows. "The Hills," near the church, are either the remains of a moot-hill or a small *burh*. (pp. 10, 99, 102, 103, 109, 115.)

Millbrook (201) standing on high ground, and dominating the Bedford valley, claims to be the prettiest village in the county. The church contains busts of Lord and Lady Holland by Westmacott. (p. 16.)

Milton Bryant (199), a little village 4½ m. S.E. of Woburn Sands station, has a small Norman church (St Peter) now much modernised, containing a monument by Chantrey. (p. 155.)

Northill (1292), a village and large parish 4 m. W.N.W. of Biggleswade, with several hamlets, has a fine dark red sandstone church rebuilt about 1400. The hamlet of Ickwell, built round a green with a maypole in the centre, is unusual. (pp. 96, 111.)

Oakley (330), a village on the Ouse 4 m. N.W. of Bedford. The church of St Mary is mainly E.E. The rood-screen is now in the aisle as the organ loft. Pillow lace is still made. (pp. 23, 113, 117.)

Odell (252), a village 9 m. N.W. of Bedford, on the Ouse, has a good Perp. church (All Saints') with the rood-screen still existing, and a fine Jacobean pulpit, with hour-glass, and some good glass. Of Odell Castle considerable traces remain. (pp. 67, 99, 117, 127, 149.)

Pavenham (308), in a bend of the Ouse 5 m. N.W. of Bedford, lies in a hilly and wooded district. The church (St Peter) though small is interesting, with a broach spire, a widening nave and transept, and much carved oak. Mat and basket making and rush-plaiting are carried on. (pp. 67, 94.)

Pertenhall (237), on the N. border of the county 3 m. S. of Kimbolton, has an interesting broach-spired church, with Trans. Norman arches in nave, a rich rood-screen, and monuments. (pp. 108, 113, 115.)

Poddington (461), a village 3 m. S. of Irchester, has a fine church (St Mary) with Norman and E.E. nave arcade, an E.E. tower with rich octagonal E.E. spire, and a Norman font. Roman remains have been found in the neighbourhood. (pp. 13, 22, 99, 103, 117.)

Potton (2156), an ancient market town on the eastern border on the L. and N.W.R. to Cambridge, a market gardening centre. There are engineering works, fell-mongering and parchment industries, mills, and a brewery. The church has a Norman font. (pp. 7, 11, 16, 32, 67, 73, 138.)

Renhold (396), a pretty village 3 m. N.E. of Bedford with a church (All Saints') containing some monuments, and an altar tomb with brasses. On high ground above the Ouse are some Danish (?) earthworks.

Ridgmont (Ridgmount or Rougemont) (540), a village 11 m. S.S.W. of Bedford, where is an experimental fruit farm. By Brogborough farmhouse (17th cent.) are earthworks. (pp. 15, 16, 51, 99, 115.)

Roxton (396), a village 4½ m. S.W. from St Neots, near the junction of the Ivel with the Ouse, the vicarage annexed to Great Barford. The church (St Mary) is Dec. and Perp. and has an altar tomb to Roger Hunt, Speaker in 1420. Chawston and Colesden are hamlets.

Salford (136), a small village 2 m. N. of Woburn Sands station on the Bucks border, with a very interesting Early Dec. church, with an open belfry of oak placed outside the west end, and three altar tombs. (p. 111.)

Sandy (3377), a very large village, 3 m. N. of Biggleswade at the junction of the G.N.R. and L. and N.W.R., on the Ivel, situated under the range of sand hills. Owing to its soil and railway conveniences it has become a great market and other gardening centre. It is not now thought to be the Roman Salinae, but "Galley Hill" and "Chesterfield" are undoubtedly the remains of Roman Camps. "Caesar's Camp" was probably a British stronghold, and the earthworks at Sandy Place possibly Danish. Many Roman relics have at different times been found. The cruciform church of St Swithin has a monument to Captain Peel. Beeston, Girtford, Seddington, and Stratford are hamlets. (pp. 6, 11, 12, 18, 44, 53, 68, 69, 70, 95, 96, 131, 132, 133, 138.)

Sharnbrook (755), a large and pretty village on the Ouse, with a station on the Midland line, 8 m. N.W. of Bedford. The church is Early Dec. and Perp. Pillow lace is made. Roman remains have been found in the neighbourhood. (pp. 11, 17, 21, 22, 148.)

Shefford (842), a small market town on the Ivel 6 m. S.W. of Biggleswade on the Midland Ry., with very wide streets. Roman relics of unusual interest have been found here and at Stanfordbury near by. (pp. 15, 16, 34, 53, 60, 96, 97, 98.)

Shillington (1588), a large village 5 m. N.W. of Hitchin ; part of the parish is in Herts' administration. The church (All Saints') stands on high ground and is one of the best in the county—an early Dec. building, with some Trans. Norman, a fine screen, and two curious square battlemented turrets. (pp. 96, 111.)

Silsoe (561), a small parish midway between Bedford and Luton, contains the large seat Wrest Park. There are good 18th cent. gardens, but the house was rebuilt on a new site in the 19th cent., converted into a hospital, and burnt in 1916.

Southill (989), a village 3 m. S.W. of Biggleswade, is chiefly noticeable for Southill House and Park, the birthplace and also the burial place of Admiral Byng, and now the seat of the Whitbreads. Roman remains have been found in the neighbourhood. (pp. 16, 53, 163.)

Stevington (Steventon) (479), a pleasant village on the Ouse 5 m. N.W. of Bedford, with an interesting church, with a tower of which a good part is pre-Norman, an Early Dec. nave arcade, and a " Holy Well " below the churchyard. There is a good village cross. Pillow lace making and rush-plaiting are carried on. (pp. 32, 82, 114, 115.)

Stotfold (3128), a large village and parish on the eastern border of the county 2½ m. N.W. of Baldock, contains the Three Counties' Asylum. (pp. 96, 132.)

Studham (320), a small village, the most southern of the county, 4½ m. S. of Dunstable, was once partly in Herts, but since 1897 wholly in Bedfordshire. The church of St Mary has nave arcades of the early 13th cent., and a Dec. chancel. The elaborate E.E. font is one of the most interesting in the county. (pp. 9, 101, 108.)

The Font, St Mary's Church, Studham

Sutton (217), a small village 2½ m. N.E. of Biggleswade, has an E.E. and Dec. church, and a picturesque old pack-horse bridge of the 13th cent., one of the few remaining. (pp. 16, 99, 111, 150.)

Swineshead (138), a little village in the N. of the county, 3 m. S.W. of Kimbolton, formerly in Hunts. Its church, St Nicholas, has many details worthy of study—a canopied Easter sepulchre, carved stalls, etc.

" Motte " and Late Norman church tower, Thurleigh

Tempsford (431), a village on the Great North Road, at the junction of the Ouse and the Ivel; is mentioned in the Eng. Chronicle as occupied by the Danes, who probably destroyed the church in 1010. Earthworks, called the Gannocks, are probably Danish. (pp. 11, 75, 76, 77, 99, 111.)

Thurleigh (= The Leigh) (433), 7 m. N. of Bedford, has a fine church (St Peter) with a central Norman tower. Near by are moated earthworks which are probably remains of a fortified manor house. (pp. 16, 99, 103.)

Tilsworth (206), a small village 2½ m. N.W. of Dunstable, has a Dec. church with some good tombs; and a moated manor farm where the fine gateway of the now demolished ancient manor house still exists. (pp. 99, 132.)

Toddington (1948), an ancient market town 5½ m. E.N.E. of Leighton Buzzard, with a fine cruciform church with central tower, the nave arcade and part of the tower being E.E. It is specially interesting for its monuments (Peyvre, Cheyne, Wentworth). Straw plait manufacture still exists. Saxon relics have been found in the neighbourhood. (pp. 12, 53, 83, 85, 95, 96, 98, 99, 115, 119, 161.)

Totternhoe (Tottenhoe) (450), a village 2 m. W. of Dunstable on the chalk. The Perp. church has a fine oak roof with figures. The quarries of "Totternhoe stone," a clunch, have been worked for centuries for churches, and much of St Albans is made of it. It is now not much used. There are fine earthworks, which appear to have been in turn British, Saxon, and Norman. There are lime and cement works. (pp. 14, 26, 70, 71, 95, 97, 99, 117, 129.)

Turvey (841), a village on the Bucks border, 7 m. W.N.W. of Bedford. All Saints' church, which shows some pre-Norman work, a Norman font, various E.E. details, and very rich ironwork on the S. door, is however mainly of interest for its remarkable series of Mordaunt monuments. (pp. 22, 28, 109, 134.)

Warden, Old (406), a picturesque village 8 m. S.E. of Bedford, is the site of a Cistercian Abbey, of which only a few remains exist. It was the origin of the famous "Warden pear." At Quince Hill various relics have been found. (pp. 96, 117.)

Turvey Abbey

Willington (370), a village on the Ouse, 4 m. E. of Bedford. There is a fine Perp. church (St Lawrence), the remains of a moated manor house, which has a very curious ancient dovecot, and some earthworks (Danish?). Roman relics have been found. (pp. 53, 76, 97, 115, 129.)

Woburn (1122), a market town on the Bucks border, 6 m. N. of Leighton Buzzard, has many attractive old houses. Woburn Abbey, the seat of the Duke of Bedford, was a Cistercian

Church, Dovehouse, and Stables, Willington

house, of which no remains exist. The present mansion was built in 1744, and contains a notable collection of pictures and sculpture; the park is one of the largest in England and is celebrated for the very fine zoological collection which has been brought together by the present owner. (pp. 7, 15, 35, 37, 49, 50, 51, 58, 60, 72, 84, 85, 86, 96, 108, 125, 148, 163.)

Woburn Sands, some 7½ m. N. of Leighton Buzzard, is a recently-created ecclesiastical parish made up of the civil parish of Aspley Heath and parts of Aspley Guise and Wavendon. Its pine-woods, sandy soil, and good air have made it a well-known health resort. (p. 12.)

Wootton (1394), a large straggling village and parish 4½ m. S.W. of Bedford, with an interesting E.E. church (St Mary) with a good chancel screen and monuments to the Monoux family.

Wymington (493), a small village at the extreme N.W. of the county on the Northamptonshire border, 13 m. N.W. of Bedford. Its church (St Lawrence) is, with that of Yielden, the best example of Dec. in Bedfordshire, with a fine spire, good brasses, and two octagonal turrets at E. end, like those of Shillington. It was entirely rebuilt by John Curteis, Mayor of the Wool Staple at Calais, who died in 1391. (pp. 13, 22, 70, 95, 111, 112, 147.)

Yielden (or Yelden) (177), a small village at the extreme N. of the county, has a very interesting Dec. church (St Mary) and the extensive remains of a castle of the Traillys. Dr Dell, Master of Caius College, the rector, incensed his parishioners by permitting "one Bunyan, a tinker" to preach in his church on Christmas Day 1659. (pp. 70, 97, 99, 111, 119, 150, 167.)

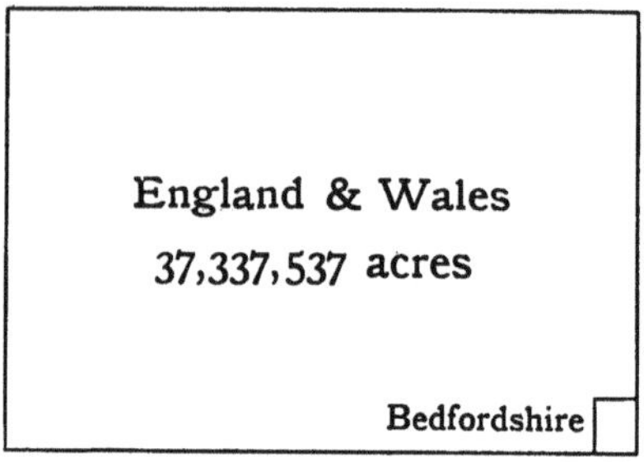

Fig. 1. Area of Bedfordshire (302,942 acres) compared
with that of England and Wales

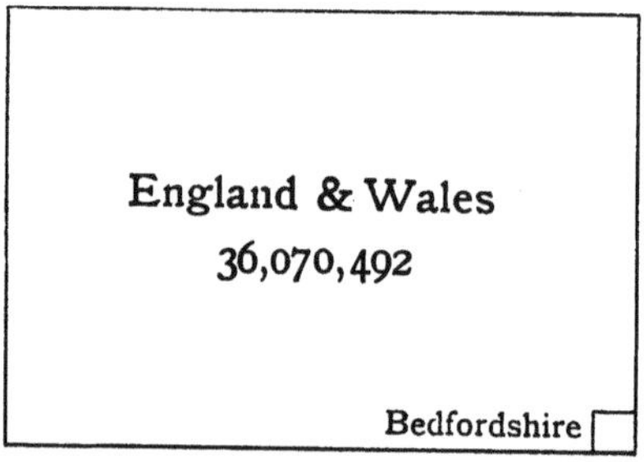

Fig. 2. Population of Bedfordshire (194,588) compared
with that of England and Wales in 1911

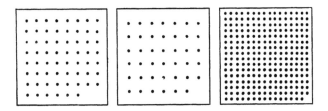

England and Wales 618   Bedfordshire 411   Lancashire 2550

Fig. 3. Comparative Density of Population to the
square mile in 1911

(*Each dot represents ten persons*)

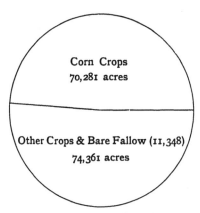

Corn Crops
70,281 acres

Other Crops & Bare Fallow (11,348)
74,361 acres

Fig. 4. Proportionate area under Corn Crops compared with
that of other cultivated land in Bedfordshire in 1913

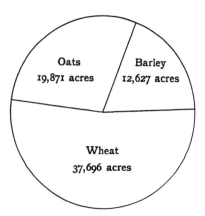

Fig. 5. Proportionate area of chief Cereals in
Bedfordshire in 1913

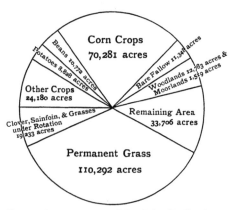

Fig. 6. Proportionate areas of land in Bedfordshire in 1913

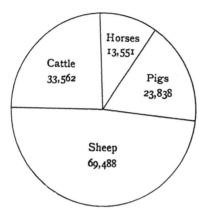

Fig. 7. Proportionate numbers of Horses, Cattle, Sheep
and Pigs in Bedfordshire in 1913

www.ingramcontent.com/pod-product-compliance
Ingram Content Group UK Ltd.
Pitfield, Milton Keynes, MK11 3LW, UK
UKHW042143280225
455719UK00001B/64

9 781107 671942